PHYLOGENETIC RELATIONSHIPS
AMONG ADVANCED SNAKES
A MOLECULAR PERSPECTIVE

Phylogenetic Relationships Among Advanced Snakes

A Molecular Perspective

by John E. Cadle

A Contribution from the Museum of Vertebrate Zoology
of the University of California at Berkeley

UNIVERSITY OF CALIFORNIA PRESS
Berkeley · Los Angeles · London

UNIVERSITY OF CALIFORNIA PUBLICATIONS IN ZOOLOGY

Editorial Board: Peter B. Moyle, James L. Patton,
Donald C. Potts, David S. Woodruff

Volume 119
Issue Date: February 1988

UNIVERSITY OF CALIFORNIA PRESS
BERKELEY AND LOS ANGELES, CALIFORNIA

UNIVERSITY OF CALIFORNIA PRESS, LTD.
LONDON, ENGLAND

ISBN 0-520-09956-7
LIBRARY OF CONGRESS CATALOG CARD NUMBER: 87-35692

© 1988 BY THE REGENTS OF THE UNIVERSITY OF CALIFORNIA
PRINTED IN THE UNITED STATES OF AMERICA

Library of Congress Cataloging-in-Publication Data
Cadle, John E.
    Phylogenetic relationships among advanced snakes.

    (University of California publications in zoology;
vol. 119)
    "Part of a doctoral dissertation to the University of
California, Berkeley"—Acknowledgments.
    Bibliography: p.
    1. Snakes—Evolution. 2. Chemical evolution.
3. Reptiles—Evolution. I. Title. II. Series:
University of California publications in zoology;
v.119.
QL666.06C18   1988   597.96'0438   87-35692
ISBN 0-520-09956-7 (alk. paper)

Contents

List of Tables, vii
List of Figures, viii
Acknowledgments, ix

GENERAL INTRODUCTION   1

MOLECULAR APPROACHES TO PHYLOGENY RECONSTRUCTION   3
    The Nature of Molecular Data and Their Use in Systematics, 3
    Molecular Clocks, 6

MATERIALS AND METHODS   8
    Biochemical and Immunological Methods, 8
    Nonreciprocity of Immunological Data, 9
    Rate Tests, 10
    Radioimmunoassay Analysis, 11
    Methods of Tree Construction from Immunological Data, 13
        General Comments, 13
        Tree Construction Methods Used in this Work, 14
    A Comment on Sampling and Approach, 14
        Taxa Included Within the Study, 14
        Approach to Molecular Analyses, 15

A MOLECULAR APPROACH TO THE PHYLOGENY OF ADVANCED SNAKES 17
    Phylogenetic Problems Within the Advanced Snakes, 17
    Results, 18
        Radioimmunoassay Analysis, 18
        Microcomplement Fixation, 20
            Reciprocal Tests, 20
            One-way tests Between *Atractaspis* and Other Advanced Snakes, 28
        Rate Tests, 29

Discussion, 33
    Monophyly of the Advanced Snakes, 33
    Phylogenetic Relationships Among Major Groups of Advanced
        Snakes, 34
    Major Lineages Within Colubrids, 35
    Implications for Evolutionary Studies of Advanced Snakes, 37

TEMPORAL FRAMEWORK FOR THE ADVANCED SNAKE RADIATION   38
    An Evaluation of the Fossil Record of Advanced Snakes, 38
        Early Caenophidians, 38
        Extant Caenophidian Families, 39
        Inferences from the Snake Fossil Record—a Critique, 41
        Conclusions, 44
    Implications of the Molecular Data, 45
    Limitations of MC'F Studies of Albumins for Resolving Relationships Among
        Advanced Snakes, 49

SUMMARY AND CONCLUSIONS   51

Appendix A: Specimens Used for the Production of Antisera, 53
    Henophidia, 53
        Boidae, 53
    Caenophidia, 53
        Viperidae, 53
        Elapidae, 54
        Atractaspis, 54
        Colubridae, 54

Appendix B: Specimens Used in Immunological Cross Reactions, 57
    Viperidae, 57
    Atractaspis, 58
    Colubridae, 58

Literature Cited, 59

List of Tables

1. Matrix of RIA^{ID} values among albumins of representatives of major groups of advanced snakes, 19
2. Matrix of MC'F immunological distances among the albumins of advanced snakes, 22
3. Matrix of corrected MC'F albumin immunological distances among advanced snakes, 23
4. Input and output values for a phylogenetic tree showing relationships among advanced snakes, 24
5. MC'F immunological distances between the albumins of *Atractaspis* and representative viperids and elapids, 30
6. Rate tests (albumin immunological distances) comparing anti-*Boa* albumin to representatives of various advanced snake families, 31
7. Rate tests comparing members of various colubrid lineages to a pool of seven elapid antisera, 32
8. Total numbers of snake genera in the North American Tertiary fossil record, 42

List of Figures

1. An example of a rate test, 11
2. Three alternative hypotheses for relationships among advanced snakes based on RIA analysis of albumins, 20
3. A Fitch-Margoliash tree based on RIA analysis of albumins, 21
4. A Fitch-Margoliash tree based on MC'F immunological distances among albumins of advanced snakes, 25
5. A Fitch-Margoliash tree based on MC'F immunological distances among albumins of advanced snakes, 26
6. A Wagner tree based on MC'F immunological distances among albumins of advanced snakes, 27
7. A consensus phylogenetic tree showing relationships among advanced snakes, 28

Acknowledgments

This study forms part of a doctoral dissertation submitted to the University of California, Berkeley. I am grateful to Dr. David B. Wake for continual guidance during my graduate study, and also to the other members of my thesis committee, Drs. Harry W. Greene and Vincent M. Sarich, for their advice and suggestions during the course of the study.

To my valued friend and colleague, Dr. James L. Patton, I owe considerable gratitude for support and guidance over the past few years. His enthusiastic assistance, intellectual stimulation, and companionship during much of the field work for this study will be forever treasured, as will his confidence in my abilities and tolerance of my way of doing things.

Dr. A. C. Wilson provided laboratory facilities during the initial phases of this project. Vincent Sarich spent many long hours teaching me the immunological methods, and I received considerable advice and assistance from Dr. Ellen Prager. For other technical assistance I am grateful to Monica Frelow and James Patton. Dr. David L. Swofford kindly provided copies of his WAGPROC and NETOPT programs and Evelyn Dilworth worked conscientiously to adapt these programs to the Berkeley computer facilities. Dr. Jerold Lowenstein graciously permitted me to use some of his radioimmunoassay data on snake relationships.

The field work for this project was accomplished with the help of numerous individuals, and special thanks are due the following: for assistance in South America, Brent Berlin, Elois Ann Berlin, James Patton, Carol Patton, John P. O'Neill, Michael D. Robinson, and Maria Alicia Barros; in California and Mexico, Samuel S. Sweet, Donald O. Straney, and Theodore Papenfuss.

Barry Hughes and Steven Spawls (Ghana), Alex Duff-McKaye and Michael Cheptumo (Kenya), and Donald G. Broadley (Zimbabwe) provided many courtesies during my stays in those countries. David and Leslie Leonard and their family, and Laurence Frank made my stay in Kenya particularly enjoyable.

I am grateful to numerous other individuals who donated specimens or samples that greatly increased the data base for this study: Stephen Busack, Herbert Dessauer, Robert Drewes, Reg Fisher, Mercedes S. Foster, J. Whitfield Gibbons, George C. Gorman, Harry Greene, James Hanken, Leo Hoevers, Elazar Kochva, Angelo Lambiris,

Carl Lieb, Roy W. McDiarmid, Robert W. Murphy, Theodore Papenfuss, James Patton, Steve Reilly, Michael Robinson, Richard Sage, Norman J. Scott, Robert Seib, H. Bradley Shaffer, Jack W. Sites, Steven Spawls, Samuel Sweet, David Wake, and Viv Wilson.

Samuel Sweet, Donald Straney, Sadie Coats, William Rainey, and Steven W. Sherwood sharpened my thinking on many critical issues. Jacques Gauthier and Drs. Joseph T. Gregory and Gareth Nelson provided stimulating and enlightening discussions on various aspects of phylogeny. For their comments on the manuscript I thank Ron Heyer, George Zug, Ron Crombie, and Charles Crumly, and I am especially grateful to Dr. Richard Estes for answering my queries concerning snake fossils and for his comments on portions of this work. Drs. Herbert C. Dessauer and Alan H. Savitzky provided detailed comments on an early version of this paper. To all of these I owe my deepest thanks, but of course any errors of fact or interpretation are my responsibility.

Research upon which this study is based received support from the Museum of Vertebrate Zoology (Annie Alexander and Louise Kellogg grants-in-aid), Sigma Xi, the National Science Foundation (BNS 76-17485 and DEB 80-14101), the Chancellor's Patent Fund, and the Tinker Foundation through the Center for Latin American Studies, University of California, Berkeley. Collecting permits were granted by the governments of Mexico, Guatemala, Peru, Venezuela, Brazil, and Zimbabwe, and their cooperation is gratefully acknowledged, as is the support of the Smithsonian Institution, where I held a postdoctoral fellowship while this manuscript was prepared. I appreciate the assistance of Carol Stone in seeing this manuscript through to publication.

Note: The author's present address is Academy of Natural Sciences, 19th and the Parkway, Philadelphia, Pennsylvania 19103.

GENERAL INTRODUCTION

Many of the traditional questions in evolutionary biology have focused on the processes promoting or constraining the diversification of lineages. The fossil record has historically supplied many insights into evolutionary changes over long periods of time and into the temporal span of living lineages. Fossils have been less generally useful in reconstructing phylogenetic relationships because the number of lineages preserved in detail as fossils is small. Thus the comparative study of living organisms plays a central role in elucidating patterns and mechanisms of phylogenesis. Recently, comparative molecular studies have stimulated the study of evolutionary radiations with reference to living organisms. By using molecular methods the phylogeny can be estimated in great detail and we can ask questions directly about evolutionary mechanisms. Molecules, therefore, are invaluable for evaluating rates and patterns of morphological evolution and in constructing or testing hypotheses of relationships among organisms.

Many of the inferences concerning the role of morphological or other constraints on the diversification of lineages will come from intensive study of those lineages which show extensive parallelism or convergence. Snakes are particularly advantageous for such studies, in that they are a highly speciose group with numerous clades showing similar adaptations correlated with major features of life history (Rabb and Marx, 1973; Savitzky, 1981, 1983). Traditional approaches to phylogenetic inference have failed to provide a robust set of hypotheses for the relationships among lineages, thus precluding further analyses of mechanisms that might be responsible for the considerable diversity among snakes.

Advanced snakes, including the vipers (pit vipers and true vipers), elapids (cobras, sea snakes, etc.), and colubrids, have proven especially difficult to analyze phylogenetically. Yet, this radiation provides an array of adaptations related to feeding and habitat utilization that are among the more complex seen in vertebrates. There are few detailed reconstructions of advanced snake phylogeny, and most of these simply recognize large assemblages of related genera, with little emphasis on the relationships among these assemblages or on intergeneric relationships within them (e.g., Underwood, 1967; Dowling and Duellman, 1978).

My work is aimed at providing a general phylogenetic framework for the advanced snake radiation within which analyses of particular evolutionary problems can be addressed. In this study I first develop, using molecular data, an hypothesis for the phylogenetic relationships among higher taxa of advanced snakes: viperids, elapids, several lineages of colubrids, and an enigmatic African genus, *Atractaspis*, of uncertain phylogenetic affinities. Resolution of these relationships is crucial to initial efforts in unraveling the evolutionary history of the venom delivery apparatus in snakes. The hypothesis developed in this part also serves as a necessary phylogenetic framework within which more detailed analyses of particular lineages of advanced snakes can proceed.

In the second part of this study the temporal framework of the advanced snake radiation is examined from the standpoint of the fossil record and the implications of the molecular phylogeny developed previously. My evaluation of the molecular data leads to conclusions concerning the evolutionary history of advanced snakes that differ from most interpretations of the fossil record. I explore possible biases inherent in the known fossil record relative to these discrepancies, and develop a model for setting limits to the temporal calibration of the molecular data. The hypotheses developed here concerning phylogenetic relationships and the evolutionary history of advanced snakes have ramifications concerning the evolution and biogeography of specific lineages within the radiation. These are discussed in a series of complementary papers (Cadle, 1984a-c, 1985).

Finally, what I perceive to be some limitations of microcomplement fixation (MC'F) studies of albumins for answering particular kinds of questions concerning the phylogeny of advanced snakes is considered.

MOLECULAR APPROACHES TO PHYLOGENY RECONSTRUCTION

THE NATURE OF MOLECULAR DATA AND THEIR USE IN SYSTEMATICS

The role of molecular data in evolutionary biology is controversial. Some authors (e.g., Throckmorton, 1978; Friday, 1980; Farris, 1981, 1985) have questioned the utility of molecular data in phylogeny reconstruction, whereas others (Sarich and Cronin, 1976; Larson et al., 1981; Wake, 1981, 1982; Fitch, 1982) have suggested that molecular data are crucial not only for the evaluation of phylogenetic relationships but also for meaningful analysis of particular evolutionary processes. Three related considerations are involved in these discussions: (1) the ability of molecular data to resolve evolutionary relationships among taxa; (2) the utility of molecular data in estimating the temporal dimensions of a given phylogeny; and (3) the methods to be used in reconstructing genealogical relationships from molecular data, whether in the form of molecular sequences or indirect measures of molecular resemblance (electrophoretic data, immunological distances, DNA hybridization data, or DNA restriction endonuclease maps). I will not attempt to review in depth all of the arguments concerning the use of molecular data for these purposes (see references cited above and Wiley, 1981: 327-339, for a general review). Rather, I direct my attention to empirical observations on the nature of molecular data, their role in inferring phylogenies, and considerations of molecular clocks.

I believe molecular data can be very powerful tools for evolutionary biology, particularly in the formulation of testable phylogenetic hypotheses. I am very much a pluralist when it comes to methods for reconstructing the evolutionary history of organisms, for I do not know of a single case in which one method has led to more than a superficial understanding of phylogeny interpreted broadly (this is, admittedly, a very subjective assessment). I also believe that a pluralistic approach should discourage dogmatism concerning the results of any one methodology (e.g., cladistic analysis of morphology) and can lead to the resolution of novel evolutionary patterns and processes (e.g., Larson et al., 1981; Patton and Smith, 1981). I would not argue that molecular approaches are "the" answer to every phylogenetic question, nor should they be unconditionally accepted to the exclusion of other data. However, the contribution of molecular studies to phylogenetic analysis has been positive and invigorating for evolutionary studies and will continue to provide critical information for difficult phylogenetic problems (see Fitch, 1982). The estimation of evolutionary history is difficult at best, and no data that might provide clues to relationships should be ignored.

Numerous observations support the independence of molecular and morphological evolution (King and Wilson, 1975; Wilson et al., 1977), and this provides a key to their utility in evolutionary studies. Molecular data allow hypotheses of phylogenetic relationships to be developed and tested independently of those derived from morphology, and questions concerning morphological evolution can then be framed in terms of a phylogeny derived independently of morphological features. These attributes of molecular data make them particularly attractive as adjuncts to studies of the evolution of morphology (see Wake, 1981, 1982; Larson et al., 1981; Sage et al., 1984).

Clearly, if molecular phylogenies never conflicted among themselves or with morphologically-based phylogenies, there would be few disagreements over their application to phylogenetic studies. Such conflicts do occur, but this is unique neither to molecular phylogenetic hypotheses nor to particular methods of phylogenetic inference. Phylogenetic hypotheses, no matter what the basic data, must be tested with reference to other comparative evidence (other molecules, morphology, chromosomes, etc.). Epistemological arguments do not constitute tests.

Most workers who have criticized the use of molecular methods in phylogenetic reconstruction have avoided testing the molecular hypotheses directly (an exception is Romero-Herrera et al., 1978). Instead, the criticisms often concern the methodologies used to derive the phylogenies (Farris et al., 1979; Farris, 1981; Friday, 1980; Swofford, 1981) or uncertainties regarding the nature of evolution at the molecular level (Throckmorton, 1978; Friday, 1980; Farris et al., 1979). These epistemological arguments require some assumptions about what molecular data are or should be, and are not tests of the hypotheses derived from the data. In fact, few molecular phylogenies have been shown to be grossly discordant with other comparative data. Numerous studies indicate, however, a congruence of molecular phylogenies both among themselves and with reference to morphological or chromosomal phylogenies (Sarich, 1969; Cronin and Meikle, 1979; Scanlan et al., 1980; Wake et al., 1978; Maxson et al., 1979; Larson et al., 1981; Yang and Patton, 1981; Prager and Wilson, 1976; Bruce and Ayala, 1979; Gorman et al., 1980; Densmore 1981; Beverly and Wilson, 1982; Shochat and Dessauer, 1981; Sibley and Ahlquist, 1984). Of course, conflicting hypotheses do develop with molecular data, as with any other data (e.g., Romero-Herrera et al., 1978). Such conflicts are inevitable in any real data sets used for phylogenetic reconstructions, but they do not obviate their use for these purposes. Rather, they are an indication that multiple criteria should be used in phylogeny reconstructions in order to better judge those hypotheses that are robust.

A second criticism concerning some molecular approaches to systematics is that immunological, genetic distance, and DNA annealing data are inherently less valuable than direct sequence date (DNA, protein) for phylogenetic analyses (Friday, 1980; Farris, 1981, 1985). Indeed, Farris (1981: 20-21) stated that, because immunological distances are non-metric, the claim that they correspond to sequence differences is unjustified. However, there are empirical data to substantiate the claim that immunological distances estimate sequence differences among proteins. A good correlation between immunological distances and sequence differences or indices of amino acid composition differences has been demonstrated for a number of proteins (Prager and Wilson, 1971a, b; Prager et al., 1978; Champion et al., 1975; Margoliash et al., 1970; Ibrahimi et al., 1979; Jolles et al., 1976, 1979; Wallace and Wilson, 1972; Arnheim et al., 1969; White et al., 1978). Benjamin et al. (1984) summarized evidence for the sequence-immunology correlation for lysozyme c, ribonuclease, myoglobin, cytochrome c, azurin, and serum albumin, and reported least-squares correlation coefficients of 0.85 to 0.96 between the sequence differences and immunological distances. These observations suggest a strong correspondence between the direct and indirect assessments of molecular evolution.

There is perhaps a bit of misplaced confidence in the relative power of sequence data for resolving phylogenies over that of some indirect measures (e.g., Farris, 1981). This is apparent from two observations concerning the large body of sequence data available. First, phylogenetic reconstructions using sequence data do not lead to unambiguous phylogenetic hypotheses: several more or less equally parsimonious topologies are possible for all data sets examined to date (e.g., hemoglobins: Beard and Goodman, 1976; Goodman et al., 1979; Sarich and Cronin, 1980; Baba et al., 1982; and Fitch, 1979; myoglobins: Romero-Herrera et al., 1978; Dene et al., 1982; Maeda and Fitch, 1981; cytochrome c: Fitch and Margoliash, 1967; lens α-crystallin: De Jong and Goodman, 1982). There are usually many alternative trees that do not differ substantially by the usual criteria for evaluating such trees (this is also true of "indirect" molecular data sets), and the best trees by some goodness of fit measure may be discordant with a well-established phylogeny based on other evidence (Fitch, 1979; Maeda and Fitch, 1981; Wyss et al., 1987). For some phylogenetic problems, abundant sequence data have proven to be as inconclusive as morphological data (Wyss et al., 1987).

Second, for primates, in which the broadest array of comparative molecular data is available (immunological: Sarich and Cronin, 1976 and included references; electrophoretic: Bruce and Ayala, 1979; mitochondrial DNA maps: Ferris et al., 1981; DNA-DNA hybridization: Sibley and Ahlquist, 1984) and in which several proteins have thus far been sequenced (Baba et al., 1980), the sequence data have had relatively little impact on the phylogenetic hypotheses developed. Sarich and Cronin (1976:143-144) observe: ". . . their [sequence data] contribution to systematics generally and to knowledge of primate evolution specifically has been at best supplemental or confirmatory. There is probably not a single case for the primates where sequence data have told us something that we did not already know about the phylogeny involved." It is unlikely, therefore, that sequence data, were they to become available for most organisms, would substantially improve the resolving power of the phylogenetic hypotheses developed by less direct molecular means. Although we should strive to derive robust phylogenies from those sequences available, abandoning other forms of molecular analysis would considerably hamper phylogenetic studies for the vast majority of organisms.

Farris (1981) provided many other critiques concerning the use of distance data in phylogenetic analysis. These mostly concern the logical basis underlying distance methods and the resulting influence on the interpretation of distances within a phylogenetic framework. I will not deal with these critiques here, but an alternative viewpoint and extended discussion of them can be found in Felsenstein (1982, 1984), some of whose points are rebutted by Farris (1985).

MOLECULAR CLOCKS

Much of the controversy concerning molecular approaches to systematics has involved not so much the actual phylogenies produced, but the concept of molecular clocks and time scales for the evolution of particular groups (see Sarich, 1973; Radinsky, 1978; Carlson et al., 1978; Wilson et al., 1977, for reviews). Two interrelated problems exist: the actual existence of molecular clocks, and the calibration of such clocks, if they do exist. Wilson et al. (1977) reviewed evidence for the existence of molecular clocks and emphasized their stochastic nature (see also Sarich, 1973; Sarich and Cronin, 1976). I wish simply to address several points which have emerged recently.

Throckmorton (1978) comments that the rate of molecular evolution through long periods of evolutionary time is unpredictable, and that molecular evolutionary data therefore can not be expected to produce reliable phylogenetic hypotheses or to be useful in dating evolutionary events. This argument apparently stems from the misconception that clocklike behavior is assumed in molecular phylogenetic analyses (see also Farris, 1981:4), but such is not the case (Sarich and Cronin, 1976). In fact, much work has been devoted specifically to testing whether molecular evolution proceeds in a clocklike fashion (Sarich and Wilson, 1967a,b, 1973; Wilson and Sarich, 1969; Sarich, 1969, 1973; Sarich and Cronin, 1976, 1980; Cronin and Sarich, 1980; Wilson et al., 1977; Fitch and Langley, 1976; Langley and Fitch, 1974; Fitch, 1976, 1982). Sarich and Cronin (1976) rightly emphasize that questions concerning the existence of molecular clocks and of their precision can only be answered empirically, and must be answered for each specific case if the data are to be used in dating evolutionary events. Thus, regularity of molecular evolution must be demonstrated for each particular protein in each group of organisms before using molecular results to date evolutionary events.

Farris (1981) questioned the existence of molecular clocks because some distance measures (particularly immunological) are nonmetric (a mathematical property), and on the contention that such distances "cannot be truly clocklike" (p. 22). Yet the observation of a stochastically regular molecular clock follows not only from distance measures but also from sequence data (Langley and Fitch, 1974; Fitch and Langley, 1976; Fitch, 1976, 1982; Wright, 1978). Moreover, observations of regularity in molecular evolution supplied by distance measures do not depend on apportioning those distances into a phylogeny (as assumed by Farris, 1981:4). Relative rate tests (Sarich and Wilson, 1967a; Sarich, 1973; Wilson et al., 1977) using outside reference species to a particular group (determined from nonmolecular data), can provide information on whether particular lineages are slower or faster than the average rate shown by members of that group.

Numerous such examples of lineage-specific rate differences have been reported: among the primates, *Aotus* albumin and *Phaner* transferrin (both slow); and *Phaner* albumin, the common anthropoid albumin lineage, the common lorisiform transferrin, and primate cytochromes *c* in general (all fast); among bats, *Rhinopoma* albumin (slow) and phyllostomatoid albumins (fast); among marsupials, *Caluromys* and *Marmosa* albumins (slow); among snakes, viperid albumins (slow) (Moore et al., 1976; Sarich and Cronin, 1976; Cronin and Sarich, 1980; Holmquist et al., 1976; Maxson et al., 1975; Dessauer et al., 1987). The major point is that these rates can be assessed for each lineage independently of other data and can be accounted for in considerations of evolutionary time for those lineages.

I conclude that the weight of empirical evidence favors the largely time-dependent nature of molecular evolution (see also Selander, 1982; Fitch, 1982) and, more importantly, that the observed regularity can be empirically assessed. This leaves a gap between some epistemological and theoretical arguments (Throckmorton, 1978; Farris, 1981, 1985) and the empirical data, and it leaves unanswered questions as to the calibration of a molecular clock for a particular group. This matter will be of direct concern later in this study.

MATERIALS AND METHODS

BIOCHEMICAL AND IMMUNOLOGICAL METHODS

Snakes were obtained in the field and from other sources; blood and/or tissue samples were preserved by several means. Most field samples were frozen in liquid nitrogen and transported to the laboratory where they were stored at -70° C. Some field samples, particularly those from Africa and Brazil, were preserved in 2-phenoxyethanol phosphate sucrose (PPS - 85.6g sucrose, 16.7 ml 1M KH_2PO_4, 83.3 ml 1M K_2HPO_4, 15 ml phenoxyethanol, made up to one liter with distilled water; Nakanishi et al., 1969) in the following proportions: for blood, 1 part PPS to 1 part whole blood or plasma; for tissue samples (mixture of heart liver, kidney, and/or muscle), 4-5 parts PPS to 1 part finely chopped tissue. PPS-preserved samples were maintained at ambient temperature until transported to the laboratory, where they were frozen at -10° C. Appendix A lists locality data and voucher specimens for those samples used in antiserum production, and Appendix B lists the samples used in immunological cross-reactions.

Albumin was initially isolated from the plasma of reference species by a one-step vertical slab polyacrylamide gel electrophoresis procedure (Cadle and Sarich, 1981). Some of the initial antisera to albumins prepared in this manner showed minor reactive components other than albumin when tested in Ouchterlony immunodiffusion tests against whole plasma. Subsequently, most or all of this heterogeneity was removed by treating plasma or tissue samples with the following rivanol procedure prior to electrophoresis: 2-3 ml plasma (more for whole blood or tissue extracts) were centrifuged for 10 min. at 15,000 RPM and the supernatant (1) was dialyzed for 6 hours at 4° C against a 1/20 dilution of Tris EDTA borate buffer, pH 8.4 (2M Tris (hydroxymethyl) aminomethane, 0.2 M boric acid, 0.032 M EDTA, to pH 8.4 with HCl). A solution of rivanol (6,9-diamino-2- ethoxy-acridine lactate) was prepared by adding 50 mg rivanol to 13 ml of the same buffer dilution. This solution was added to the dialyzed supernatant (1) from above, resulting in the precipitation of most of the serum albumin, along with other serum components. After 30 minutes this mixture was centrifuged for 10 min. at 15,000 RPM. The supernatant (2) was poured off and saved. The precipitate was allowed to break up overnight in 0.5 M Tris HCl (TRIZMA HCl, Sigma no. T-3253) buffer, releasing albumin into solution and freeing the rivanol; the presence of albumin in this solution was checked by immunodiffusion tests against albumin antisera already prepared. The mixture was again centrifuged, and the supernatant (containing albumin) was dialyzed for several hours against a 1/10 dilution of Tris sulfate pH 9 buffer and then vacuum-dialyzed down to approximately 2 ml. The albumin was further purified by polyacrylamide gel electrophoresis (PAGE) as described (Cadle and Sarich, 1981).

Antisera were prepared in rabbits by the following procedure: 1.5 ml albumin solution from the PAGE were mixed with 1.8 ml Freund's Complete adjuvant, and 1 ml of the resulting emulsion was injected intradermally into two back sites on each rabbit. At 8 weeks a similar injection was made using Freund's Incomplete adjuvant. These were followed on the 11th and 12th weeks with an intravenous injection of 1 ml albumin solution per rabbit. Rabbits were bled one week following the last injection, the plasma separated by centrifugation, and the antisera stored at -10° C. Individual antisera were titered using the microcomplement fixation method (MC'F) and pooled in inverse proportion to their MC'F titers (see Champion et al., 1974; Levine, 1978). All cross reactions were carried out with these pooled antisera. Antisera from three rabbits were used for making most antiserum pools, but a few were made from only two antisera.

MC'F experiments were performed as in Champion et al. (1974; further discussed in Levine, 1978), and the results were expressed in terms of albumin immunological distances (AID). Immmunological distance is defined as $100 \log_{10} I.D.$, where I.D. is the Index of Dissimilarity, the factor by which the antiserum concentration must be raised for a heterologous antigen to give a MC'F peak equal in amplitude to that given by the homologous antigen.

NONRECIPROCITY OF IMMUNOLOGICAL DATA

The rationale and the basic approach in reconstructing phylogenies from molecular data are discussed by Sarich (1973) and Sarich and Cronin (1976). They consider problems inherent in the use of immunological methods, of which nonreciprocity is of particular concern here. Nonreciprocity is the property that, for species A and B, the immunological distance between them measured with an antiserum to A will usually not equal that measured with an antiserum to B. Sarich and Cronin (1976) define "percent nonreciprocity" as

$$\% \text{ nonreciprocity} = 100 \times \left(\frac{(\text{anti-A with B}) - (\text{anti-B with A})}{(\text{anti-A with B}) + (\text{anti-B with A})} \right)$$

and I use that value as an index of the "noise" inherent in reciprocal measurements of immunological distance values by MC'F. Another measure of nonreciprocity (percent standard deviation from reciprocity) was suggested by Maxson and Wilson (1975).

One problem inherent in immunological approaches to phylogenetic inference is that nonreciprocities are often distributed nonrandomly among antisera. That is, when numerous reciprocal comparisons are available, it is generally observed that particular antisera may consistently give either higher or lower average immunological distances than do reciprocal tests (Maxson et al., 1975; Sarich and Cronin, 1976). The degree of nonrandomness can be quantified by noting, for example, the ratio of row to column sums in a distance matrix (see Table 2 of this work; Sarich and Cronin, 1976, 1980), and this value ranges up to about 50% in those data sets examined to date. The extent of this bias can then be taken into account relative to how much confidence one might have in the precise cladistic placement of a given taxon.

Sarich and Cronin (1976) describe a method for reducing this nonrandom element (see also Post and Uzzell, 1981). Briefly, this involves calculating row to column sums for a given matrix, then multiplying each value in a particular data column by the corresponding row to column ratio, starting with the most discrepant. Corrected row sums are then recalculated after each step and the process is repeated until row and column sums agree closely. For many data sets this procedure will make no difference in the actual cladistic conclusions drawn (e.g., the data in Table 2). However, this feature of immunological data must be recognized, for example, (1) when interpreting data for which only one-way comparisons to a variety of antisera are available for a given taxon; (2) when evaluating inferences concerning the temporal dimensions of a phylogeny (Maxson et al., 1975); and (3) when assessing how reliable, in molecular terms, any cladistic conclusions drawn from reciprocal immunological data might be (Sarich and Cronin, 1980).

RATE TESTS

In order to draw firm phylogenetic conclusions from immunological distance data, the amount of albumin change along lineages within a group needs to be estimated with reference to an outside group (see Sarich, 1973; Sarich and Cronin, 1976). These relative rate tests (described in more detail by Sarich, 1973; Wilson et al., 1977) help avoid errors in phylogenetic inference such as the association of taxa because of lack of molecular change in one or more of them. A simple example is illustrated in Fig. 1, where using only the relative distances among taxa A, B, and C might lead to the relationships depicted in 1b. Use of a known outside reference species (X), however, indicates that A has changed relatively more from a common ancestor than has either B or C, and, thus, that the distances among A, B, and C should be apportioned as in 1a. These tests are especially important in the placement of taxa for which only one-way immunological comparisons are available, and they are necessary in rooting molecular phylogenetic trees.

In practice, the selection of outgroups should be made judiciously, using well-established phylogenetic relationships where possible. This is a problem for advanced snakes, and my approach has been to assume first that they are a monophyletic group relative to primitive snakes, as represented by *Boa* (see Rieppel, 1979; Groombridge, 1979; and the radioimmunoassay data discussed below). The advanced snake phylogenetic tree is then first rooted using *Boa* as an outside reference. Once that pattern of relationships is established, it becomes possible, for example, to use elapids as an outgroup for apportioning amounts of albumin change among colubrid lineages or vice versa.

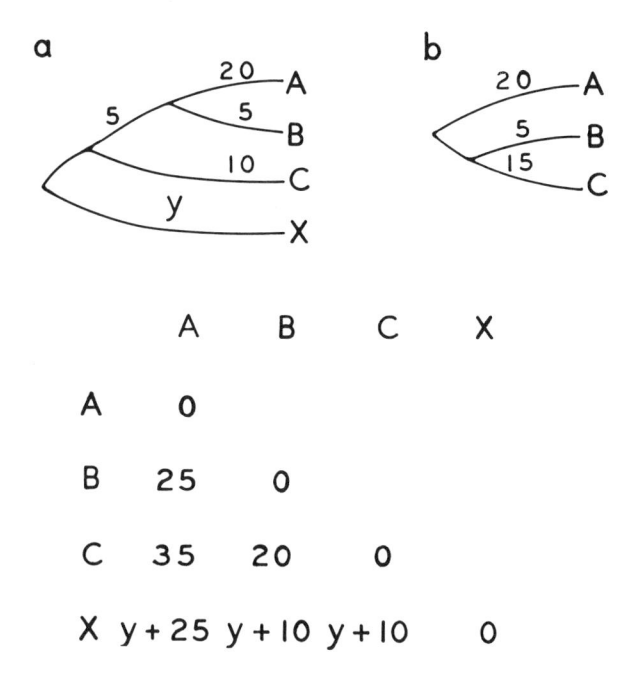

```
         a                              b
              20                              20
            5    A                         5    A
          5   5  B                           5  B
              10 C                           15 C
            y                                   
                 X
```

	A	B	C	X
A	0			
B	25	0		
C	35	20	0	
X	y + 25	y + 10	y + 10	0

FIG. 1. An example of a rate test. The observed distances are given in the matrix. Without the use of outside reference group X, the distances among A, B, and C could be apportioned as in 1b. The rate tests to X indicate, however, that A has changed 15 units more than have either B or C since they last shared a common ancestor. The immunological distances are therefore apportioned as in 1a to account for these rate differences among lineages.

RADIOIMMUNOASSAY ANALYSIS

Some major questions about snake phylogeny concern the relationships among higher taxa within the advanced snakes. To properly analyze these relationships using molecular data, outgroups (in this case primitive snakes) must be included in order to root the phylogenetic tree. Here there are problems with data acquisition by the standard MC'F procedure, because distances between boids and a sampling of elapid and colubrid taxa approach or exceed the limits of resolution of MC'F, with immunological distances generally greater than 150 (see below). Thus, obtaining reciprocal comparisons with a broad array of advanced snakes becomes technically difficult and time-consuming, and still leaves questions as to the reliability of these results in inferring relationships. Consequently, my MC'F data set for such comparisons is limited. In addition, albumin immunological distances within elapids and colubrids are not apportioned consistently using anti-*Boa* albumin, thereby suggesting that *Boa* may be too distant to serve as an appropriate outgroup.

Radioimmunoassay (RIA; Lowenstein, 1980a,b, 1981; Lowenstein et al., 1981) provides a sensitive means of addressing these broader systematic questions. This method has aided in analysis of the phylogenetic relationships among fossil and living taxa, and the data show a direct relationship to MC'F immunological distances (Lowenstein et al., 1981). Dr. Jerold Lowenstein and I are investigating the relationships among higher taxa of snakes using this method, and a preliminary data matrix used to assess the relationships among higher taxa of advanced snakes is presented here. More extensive comparisons and analysis will be reported elsewhere.

In the RIA tests reported here we used pooled antisera prepared for the MC'F tests, and the results are based on an inhibition (competitive binding) technique. Details of the experimental procedure can be found in Lowenstein (1980a) and Lowenstein et al. (1981); a brief description follows.

The standard RIA is a double-antibody system using polyvinyl microtiter plates as a solid-phase ligand. Serum dilutions are left briefly in the microtiter cups, where some protein binds to the plastic. The remainder is washed out with a solution of chicken egg protein to saturate the plastic. Rabbit antisera to albumins are then added and left for 24 hours, during which the antibody binds to the plastic-bound albumin; the rest of the antiserum is washed out. Next ^{125}I-labeled goat anti-rabbit gamma globulin is added to each cup, and it binds to the rabbit antibody present. The unbound gamma globulin is then washed out, and radioactivity is measured in each cup by scintillation counting.

Competitive binding techniques improve the specificity of the RIA. Rabbit antialbumins are first mixed with excess heterologous sera, which bind essentially all of the antibody directed against common determinants in the two species. Those antibodies reacting more strongly with the determinants on the plastic-bound albumins bind to the latter. Thus, the more similar the two competing albumins, the less will be the uptake of antibody on the plastic. Results are recorded as inhibition similarities (RIA^{IS}), which are the amounts of inhibition shown by heterologous sera relative to that of the homologous sera. Inhibition difference (RIA^{ID}) values are then calculated as $ID = 100 \log_{10} IS$.

For the RIA results reported here, representative samples of elapids, colubrids, viperids, and *Atractaspis* were used, and antisera to *Boa* and chicken albumins were used as outgroups to these.

METHODS OF TREE CONSTRUCTION FROM IMMUNOLOGICAL DATA

General Comments

Several methods have been employed to construct hypotheses of phylogenetic relationships from immunological data. The most widely used are the distance Wagner procedure (Farris 1972), the Fitch and Margoliash method (1967), and the additive algorithm of Sarich and Cronin (1976). The latter two methods are operationally very similar. Swofford (1981) presents a modification of Farris's Wagner method and describes a method for optimizing branch lengths so that, for a given tree topology, the goodness of fit criterion of Prager and Wilson (1976; the F-value) is minimized (other goodness of fit criteria can also be minimized; see Swofford, 1982a).

Since the various methods of tree construction can result in different evolutionary trees constructed from the same data matrices, some criterion must be adopted in evaluating these alternative hypotheses. The one adopted in this work, and by most systematists, is goodness of fit, or how well the constructed tree distances correspond to the input data. Several measures of goodness of fit are available: percent standard deviation (Fitch and Margoliash, 1967), f-value (Farris, 1972), and F-value (Prager and Wilson, 1976) are examples. The assumptions of the distance Wagner procedure and the Fitch-Margoliash method are summarized in Swofford (1981) and Felsenstein (1982). I will comment on only two practical aspects of tree construction using these methods.

There has been much discussion (e.g., Farris et al., 1979; Farris, 1981; Swofford, 1981) about the failure of some distance measures to satisfy the triangle inequality (i.e., to be nonmetric) and about the production of negative branch lengths in trees by some algorithms (e.g., Fitch and Margoliash, 1967). I commented above on the nonmetric property in reference to molecular clocks. Here it is of relevance only in that some authors (e.g., Farris, 1981) suggest that this property precludes the use of immunological distances in tree construction, at least using methods currently available. Thus, whatever phylogenetic content there may be in immunological data is said to be unobtainable; empirical evidence reviewed above suggests otherwise. Felsenstein (1982) states that it "is not obvious why this [nonmetricity] should be considered a fatal objection on any grounds other than aesthetic ones." He points out the need to examine the statistical properties of the various tree building algorithms with attention to assumptions underlying each. This is not currently possible with most of the algorithms available, and, except for the goodness of fit criterion, evaluation of results must be made with reference to criteria external to the actual phylogeny produced (i.e., empirical tests using other comparative data; see above). See Felsenstein (1982) for further discussion.

Similar arguments apply to the appearance of negative branches in some phylogenetic trees. These quite commonly appear in Fitch-Margoliash trees, but also occasionally in Wagner trees, though this is not widely acknowledged. It is a simple matter to restrict the Fitch-Margoliash algorithm to avoid negative branches (Felsenstein, 1982) if this is considered desirable; likewise, Swofford (1981) has developed a procedure that allows negative branches in Wagner trees. In contrast to reliance on a single, all-purpose method for deriving evolutionary trees (Farris, 1981; Swofford, 1981), the use of several methods permits evaluation of more alternatives which reflect different assumptions about the data or about the underlying evolutionary process. This approach discourages the justification of numerical phylogenetic methods on the basis of their geometric or aesthetic properties, and should discourage dogmatism (Felsenstein, 1982; see also Peacock, 1981). Until we have methods of testing alternative trees statistically, an eclectic approach would seem to be a rational one.

Tree Construction Methods Used in this Work

Matrices of immunological distances were converted into phylogenetic hypotheses using the following programs implemented on the Berkeley CDC 6400 or IBM 4341: EVOLVE (by W. M. Fitch, based on the algorithm of Fitch and Margoliash, 1967); WAGPROC (by D. L. Swofford, based on the distance Wagner procedure of Farris, 1972, as modified by Swofford, 1981; see Swofford, 1982a); and NETOPT (by D. L. Swofford, with options for fitting branch lengths of a constructed tree to an associated distance matrix in order to optimize several goodness of fit criteria; see Swofford, 1982b). I prefer Swofford's (1981) distance Wagner procedure over the original version (Farris, 1972; implemented as the WAGNER-78 program) because it includes several options for adding taxa to a partially constructed network and therefore leads to multiple solutions for a given data matrix, rather than a unique solution as in the WAGNER-78 program. Farris has subsequently developed a Wagner algorithm that produces multiple solutions, but that method has not been available to me. In all cases where WAGPROC was used, I used the Multiple Addition Criterion (Swofford, 1981). I did not construct many trees using Sarich and Cronin's (1976) method, although the results obtained with this method were, in general, very similar to those from the Fitch-Margoliash trees of best fit.

Clearly, it is not possible to examine all trees for a given data set in order to find the one of best fit (see Felsenstein, 1978, for a discussion of this problem). For the larger data matrices discussed in this work, I examined up to 40 trees constructed using EVOLVE and up to 7 using WAGPROC. These have, in general, included all of those trees that might be considered likely on the basis of other data, and I used both EVOLVE and NETOPT to construct other alternatives.

A COMMENT ON SAMPLING AND APPROACH

Taxa Included Within the Study

This study derives a general framework for interpreting the evolutionary history of advanced snakes from immunological studies of albumin. It will be clear that I have (necessarily) been selective in the choice of genera examined, so it is useful to indicate the breadth of higher categories represented. I use the term *advanced snakes* to refer to the Colubroidea of most recent authors (e.g., McDowell, 1975, 1987; Rieppel, 1979), but excluding *Acrochordus*, as in Dowling and Duellman (1978; see McDowell, 1975; Rieppel, 1979). Within this group, therefore, are the traditionally recognized families Viperidae, Elapidae (including the sea snakes; see below), Colubridae (sensu lato), and *Atractaspis*. Appendices A and B list the species examined in each category.

Some further comment is pertinent regarding the Colubridae. As pointed out below, a major problem is whether this large family (sensu lato) is monophyletic. My sampling, as discussed in this work comprises representative New World colubrines (a lineage including most New World snakes with nonbilobed hemipenes bearing a simple sulcus spermaticus; antisera to *Lampropeltis*, *Trimorphodon*, and *Chilomeniscus* albumins were produced) and xenodontines (New World snakes with a divided or secondarily simple sulcus spermaticus; antisera to *Alsophis*, *Clelia*, *Coniophanes*, *Eridiphas*, *Geophis*, *Helicops*, *Leptodeira*, *Sibon*, and *Xenodon* albumins were produced). Other data (Cadle, 1984c and unpublished) suggest that natricines (sensu Malnate, 1960) share a common lineage with colubrines and xenodontines relative to elapids and viperids, but it is not at all clear that numerous Old World "colubrids" (lycodontines, homalopsines, aparallactines) share such a history (see McDowell, 1987; Cadle, 1987). I use the terms *Colubridae* and *colubrid* loosely to refer to all of these groups (i.e., the Colubridae sensu lato), but recognize that future data may require modification of this concept.

The major problems addressed in this paper are the broad outlines of albumin evolution in advanced snakes, the age of major lineages within that group, and the implications that these data have for some generalizations commonly made concerning the advanced snake radiation. Phylogenetic data acquired for groups not considered here can be incorporated as they become available. I have discussed elsewhere (Cadle, 1984a-c) issues relevant to some of these problems.

Approach to Molecular Analyses

In all of the molecular analyses discussed herein, I have attempted to test as great a diversity of taxa within each category (genera, families, etc.) as possible given the available material. Nevertheless, the generalizations which result (for example, that colubrids form a monophyletic group relative to other advanced snakes) may be altered as the data base is increased.

My general approach has been first to develop a broad phylogenetic framework within which a more detailed consideration of relationships for particular groups could be placed. Thus the RIA data (Table 1) and the large microcomplement fixation data matrix (Table 2) include representatives of all major groups of advanced snakes: viperids (*Crotalus*, *Bothrops*, *Bitis*), elapids (*Micrurus*, *Laticauda*, *Hydrophis*, *Dendroaspis*), *Atractaspis*, and colubrids (all remaining genera). The RIA analysis (see below) indicates that advanced snakes are monophyletic relative to primitive snakes and that viperids are an outgroup to the remaining advanced snakes. These were then accepted as working hypotheses in analyzing the MC'F data.

Once the general pattern of relationships was established, genera were added, using further comparisons within groups, thus successively apportioning the data into manageable units. Because reciprocal comparisons nearly always permit a more precise placement of taxa in a molecular phylogeny than do one-way comparisons, I have produced antisera to many taxa in addition to those in Table 2. I discuss these data elsewhere (Cadle, 1984a-c) with reference to more precise placement of specific taxa within the phylogenetic framework developed here.

A MOLECULAR APPROACH TO THE PHYLOGENY OF ADVANCED SNAKES

PHYLOGENETIC PROBLEMS WITHIN THE ADVANCED SNAKES

The relationships among higher taxa of advanced snakes have posed some of the most difficult problems in the study of snake evolution. Resolution of these phylogenetic questions is necessary before any meaningful discussion of evolutionary mechanisms and the phylogeny of particular morphological and behavioral features within the advanced snakes can ensue. There are problems not only in how particular groups are related cladistically, but also in the number and composition of higher taxa of advanced snakes.

Most workers recognize at least three major lineages of advanced snakes: viperids, including pit vipers and true vipers; elapids; and colubrids. In addition, some (e.g., Bogert, 1943; Voris, 1977) recognize the sea snakes as another major lineage (Hydrophiidae), and other workers divide the Colubridae into a number of families (e.g., Underwood, 1967). There is now considerable evidence to support a close relationship of all sea snakes to terrestrial elapids (McDowell, 1969a; Cadle and Gorman, 1981; Coulter et al., 1981; Minton and da Costa, 1975; Mao et al., 1983; Schwaner et al., 1985), although the monophyly of the sea snakes within the Elapidae remains problematical (Cadle and Gorman, 1981). The Elapidae as discussed herein includes the sea snakes.

Most discussions of phylogenetic relationships among the advanced snakes have focused on the evolution of venom delivery systems and on the composition of the classically recognized families Viperidae, Elapidae, and Colubridae. The various phylogenetic hypotheses have been discussed at length in several recent works (Savitzky, 1978; Cadle and Sarich, 1981; Cadle, 1982b); I will note only the major points of the controversies here. The basic problem is that both viperids and elapids are unambiguously definable only on the basis of possessing solenoglyphous or proteroglyphous venom delivery systems, respectively, and the colubrids lack these features. Thus the colubrids are often viewed as the primitive stock of advanced snakes, and are assumed to be outgroups for the assessment of evolutionary trends in viperids and elapids (e.g., Marx and Rabb, 1972; Marx et al., 1982). No derived features have been identified which define the colubrids as a clade exclusive of the viperids and elapids, whereas the latter are defined by features possibly unique to those clades. The

Microcomplement Fixation

Reciprocal Tests. MC'F titers ranged from 1300 to 9500, with an average titer of 4000. The 18 X 18 matrix (Table 2) had a nonreciprocity of 9%, a very good figure for a matrix of that size (comparable values for mammalian albumins are 5-10% but may be twice that for particular data sets; Sarich and Cronin, 1976). When this matrix was corrected for the nonrandom reciprocity error (Table 3; see Materials and Methods section above for discussion of this procedure), the nonreciprocity dropped to 6.2%. These values were averaged and used for phylogenetic reconstruction.

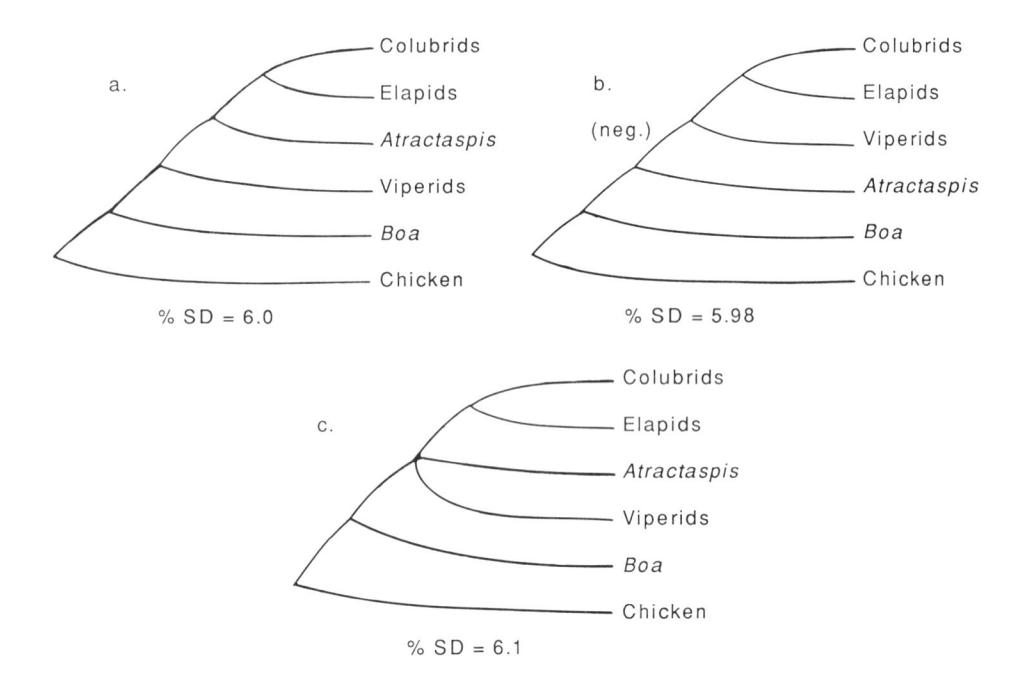

FIG. 2. Three alternative phylogenetic hypotheses for relationships among major groups of advanced snakes, as assessed by radioimmunoassay (RIA) analysis of albumins. The branching orders shown here represent the lowest %SD networks produced by optimization of Fitch-Margoliash and Wagner trees. Topology b had the lowest %SD, but introduced a negative branch.

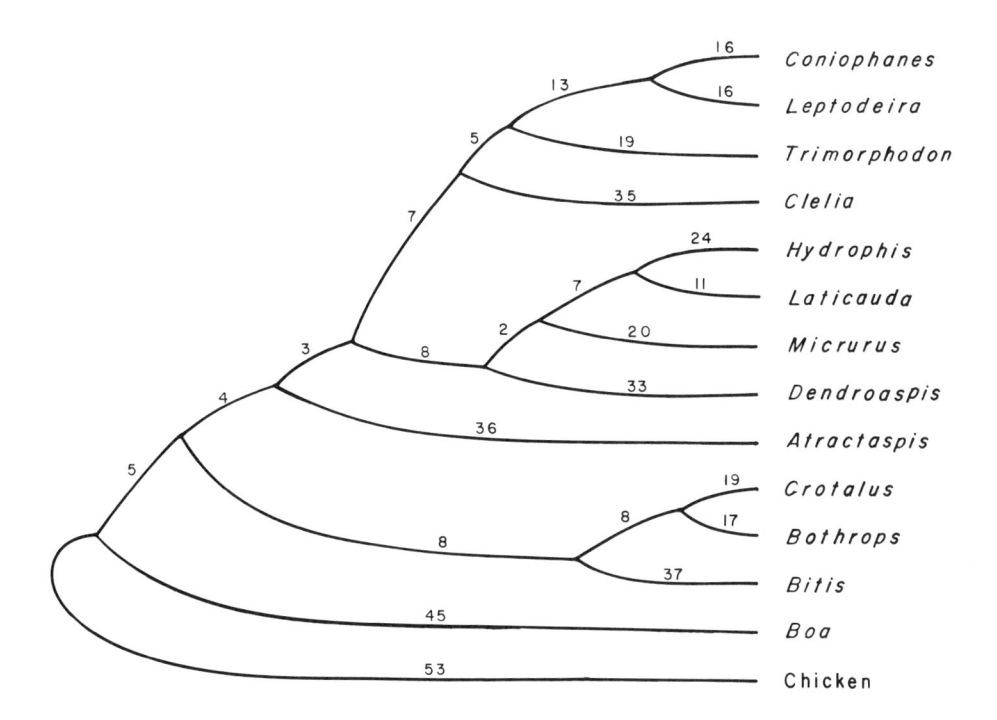

FIG. 3. A Fitch-Margoliash tree constructed from the RIA[ID] data in Table 1. The branching order corresponds to that in Fig. 2a. Here one can note the conservative albumin evolution indicated for the viperid lineage in general, and the greater conservatism of the pit vipers (*Crotalus, Bothrops*) relative to true vipers (*Bitis*) (cf. Table 6).

For analyses of these data I assumed that viperids are an outgroup to the remaining advanced snakes based on the RIA results (see Discussion below). The lowest %SD Fitch-Margoliash trees constructed from both the raw data (Table 2) and the corrected matrix (Table 3) did not differ in branching order and differed only slightly in branch lengths. Four Fitch-Margoliash trees produced initially by EVOLVE were equally good by the %SD criterion (%SD = 10.1%). One of these optimized trees is shown in Fig. 4 and the corresponding input and output half-matrices in Table 4. *Atractaspis* consistently appeared as an early derivative of the elapid lineage in these Fitch-Margoliash trees. The best Fitch-Margoliash trees differed in branching order (1) among the three major colubrid lineages and (2) among *Clelia-Alsophis-Helicops* within one of those lineages (points A and B, respectively, in Fig. 4). The Fitch-Margoliash trees based on the raw data matrix resulted in small negative branches at these points, but the optimized trees showed very small (< 1 unit) positive branches.

Table 2. Immunological distances among advanced snake taxa, obtained by using MC'F assay of serum albumins.[a]

Antigens	Cl	Tr	Co	Le	Mi	La	Cr	Hy	Xe	Lm	Si	Ge	At	Ch	Al	Er	He	Bo	Heterodon	Row Sum
Clelia	0	68	80	75	94	75	104	97	72	63	79	73	116	53	44	69	35	93	82	1290
Trimorphodon	71	0	62	81	64	69	96	77	72	31	59	65	93	13	76	96	60	75	70	1160
Coniophanes	81	66	0	36	103	71	71	99	96	74	29	14	90	40	91	42	76	77	82	1156
Leptodeira	79	70	32	0	111	75	85	84	96	68	37	24	112	59	65	32	71	72	66	1172
Micrurus	94	75	71	107	0	38	78	40	93	72	84	88	74	50	71	110	66	104	92	1315
Laticauda	88	61	75	97	43	0	83	25	72	67	81	73	72	44	64	102	63	74	81	1184
Crotalus	104	93	107	100	82	62	0	80	124	111	93	80	93	76	99	82	96	21	88	1503
Hydrophis	99	67	72	95	52	40	93	0	100	68	79	78	71	52	83	114	96	81	85	1340
Xenodon	62	60	74	70	78	58	98	105	0	72	68	64	94	63	49	84	55	91	96	1245
Lampropeltis	76	28	60	67	52	64	96	77	60	0	62	86	100	19	72	77	56	96	83	1148
Sibon	88	60	27	28	86	80	96	98	98	90	0	16	100	47	90	43	71	76	81	1194
Geophis	88	63	28	41	87	80	96	91	87	85	16	0	94	51	60	53	72	69	73	1161
Atractaspis	99	83	76	111	70	59	69	69	121	93	83	80	0	70	107	100	84	79	*	1453
Chilomeniscus	79	14	53	75	59	68	89	69	77	34	58	48	90	0	73	88	58	103	*	1135
Alsophis	46	87	90	93	63	69	99	91	57	86	78	66	103	59	0	84	34	106	95	1310
Eridiphas	78	66	38	23	89	69	76	76	96	83	41	34	90	45	68	0	60	66	65	1098
Helicops	40	78	78	62	60	47	86	73	51	74	85	52	101	50	35	63	0	88	74	1123
Bothrops	118	102	109	92	85	56	24	72	122	108	101	86	98	93	107	95	95	0	*	1123
Ninia	90	59	34	39	85	73	101	97	105	85	15	10	90	49	76	52	75	75	*	1563
Farancia	72	124	103	*	116	79	87	110	113	107	94	96	121	95	117	100	87	98	98	
Column Sum:	1390	1141	1132	1253	1278	1079	1439	1323	1494	1279	1133	1027	1591	884	1254	1334	1148	1371		
Row/Column:	0.93	1.02	1.02	0.94	1.03	1.10	1.04	1.01	0.83	0.90	1.05	1.13	0.91	1.28	1.04	0.82	0.98	1.14		

[a] Reciprocal comparisons are available for 18 genera (*Clelia-Bothrops*). One-way comparisons are available for cross-reactions involving *Ninia sebae* and *Farancia abacura*. One-way comparisons using an antiserum to *Heterodon* albumin are available for most taxa. An asterisk (*) indicates that no value is available. Row and column sums do not include one-way comparisons; the nonreciprocity for these data is 9%.

Table 3. Matrix of corrected MC'F albumin immunological distances.[a]

	Cl	Tr	Co	Le	Mi	La	Cr	Hy	Xe	Lm	Si	Ge	At	Ch	Al	Er	He	Bo
Clelia	0	68	80	70	94	82	104	97	60	57	79	82	106	68	44	58	35	105
Trimorphodon	66	0	62	75	64	75	96	77	60	28	59	73	85	17	76	80	60	85
Coniophanes	75	66	0	33	103	77	71	99	81	67	29	16	82	51	91	35	76	87
Leptodeira	73	70	32	0	111	82	85	84	81	61	37	27	102	75	65	27	71	81
Micrurus	87	75	71	100	0	41	78	40	78	65	84	99	67	64	71	91	66	117
Laticauda	82	61	75	90	43	0	83	25	60	60	81	82	66	56	64	85	63	84
Crotalus	97	93	107	93	82	68	0	80	104	100	93	90	85	97	99	68	96	24
Hydrophis	92	67	72	88	52	44	93	0	84	61	79	88	65	67	83	95	96	91
Xenodon	58	60	74	65	78	63	98	105	0	65	68	72	86	81	49	70	55	103
Lampropeltis	71	28	60	62	52	70	96	77	50	0	62	97	91	24	72	64	56	108
Sibon	82	60	27	26	86	87	96	98	82	81	0	18	91	60	90	36	71	86
Geophis	82	63	28	38	87	87	96	91	73	77	16	0	86	65	60	44	72	78
Atractaspis	92	83	76	103	70	64	69	69	102	84	83	90	0	90	107	83	84	89
Chilomeniscus	73	14	53	70	59	74	89	69	65	31	58	54	82	0	73	73	58	116
Alsophis	43	87	90	86	63	74	99	91	48	77	78	75	94	76	0	70	34	120
Eridiphas	73	66	38	21	89	75	76	76	81	75	41	38	82	58	68	0	60	75
Helicops	37	78	78	58	60	51	86	73	43	67	85	59	92	64	35	52	0	99
Bothrops	110	102	109	86	85	61	24	72	102	97	101	97	89	119	107	79	95	0

[a] Based on the row/column ratios in Table 2, and using the correction procedure described in the text.

Table 4. Input and output values for the phylogenetic tree in Figure 4.[a]

	Cl	Tr	Co	Le	Mi	La	Hy	Xe	Lm	Si	Ge	At	Ch	Al	Er	He	Bo	Cr
Clelia	0	71	77	76	82	79	88	56	70	79	78	94	70	44	76	36	102	98
Trimorphodon	67	0	66	65	70	67	76	66	29	68	67	82	16	69	65	61	90	85
Coniophanes	78	64	0	34	77	74	83	72	65	25	24	89	65	75	34	67	97	92
Leptodeira	72	73	33	0	76	73	82	71	64	37	35	88	63	74	24	66	96	91
Micrurus	91	63	87	106	0	40	48	78	70	80	78	66	69	81	76	73	81	76
Laicauda	82	68	76	86	42	0	35	75	67	77	75	63	66	78	73	70	78	73
Hydrophis	95	72	86	86	46	35	0	83	75	86	84	72	75	86	82	78	86	82
Xenodon	59	60	78	73	78	62	95	0	66	75	73	89	65	54	71	46	97	93
Lampropeltis	64	28	64	62	59	65	69	58	0	68	66	82	27	69	64	61	89	85
Sibon	81	60	28	32	85	84	89	75	72	0	17	92	67	78	37	70	100	95
Geophis	82	68	22	33	93	85	90	73	87	17	0	90	66	76	35	68	98	94
Atractaspis	99	84	79	103	69	65	67	94	88	87	88	0	81	93	88	84	93	88
Chilomeniscus	71	16	52	73	62	75	68	73	28	59	60	86	0	68	64	60	89	84
Alsophis	44	82	91	76	67	69	87	49	75	84	68	101	75	0	74	35	100	96
Eridiphas	66	73	37	24	90	80	86	76	70	39	41	83	66	69	0	66	96	92
Helicops	36	69	77	65	63	59	85	49	62	78	66	88	61	35	56	0	92	88
Bothrops	108	94	98	84	101	73	82	103	103	94	88	89	118	117	77	97	0	24
Crotalus	101	95	84	89	80	76	87	101	98	95	93	78	93	99	72	91	24	0

[a] Input values are shown below the diagonal, and the corresponding output values calculated from the tree are given above the diagonal.

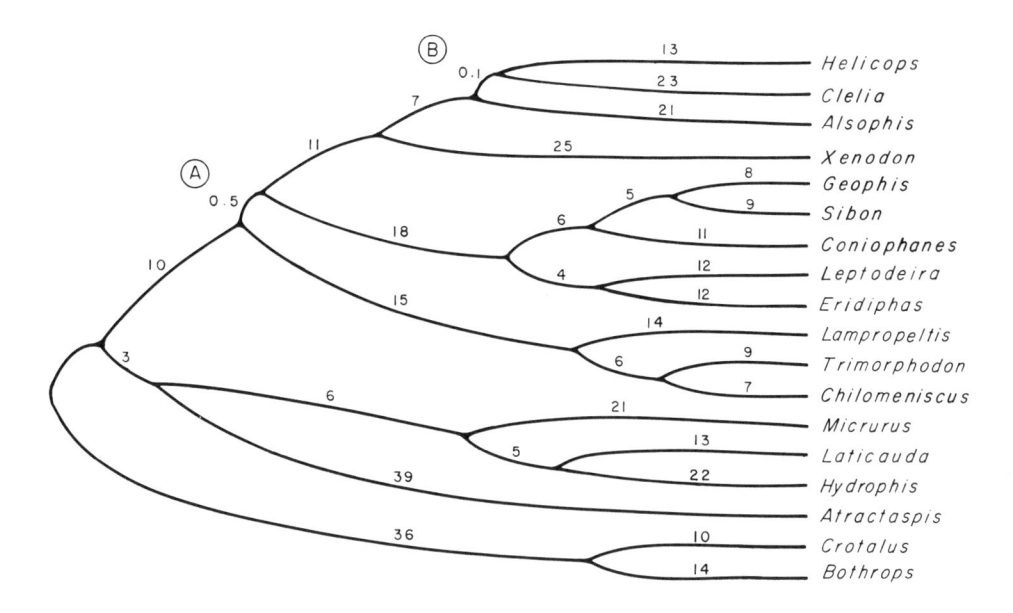

FIG. 4. A Fitch-Margoliash tree constructed from the corrected reciprocal matrix of MC'F albumin immunological distances (Table 3) and with branch lengths optimized using NETOPT. The standard deviation is 10.1%. Points A and B represent nodes at which the four "best" Fitch-Margoliash trees differed in branching order (by switching the three branches at either point). See text for further discussion. The taxonomic categories with which these genera have generally been associated are (1) xenodontine colubrids: *Helicops, Clelia, Alsophis, Xenodon, Geophis, Sibon, Coniophanes, Leptodeira, Eridiphas*; (2) colubrine colubrids: *Lampropeltis, Trimorphodon, Chilomeniscus*; (3) Elapidae: *Micrurus, Laticauda, Hydrophis*; (4) Viperidae: *Crotalus, Bothrops*. See text for discussion of *Atractaspis*.

Because *Atractaspis* was placed outside the elapid clade by the RIA data, I also constructed a Fitch-Margoliash tree showing this arrangement from the MC'F data. The resulting optimized tree (Fig. 5) has the same %SD (10.1%) as the Fitch-Margoliash trees showing *Atractaspis* as a branch of the elapid lineage.

Wagner trees constructed from both the raw data matrix and the corrected matrix had much higher %SD's than did the Fitch-Margoliash trees (see Swofford, 1981, for discussion). Optimization of the Wagner trees considerably lowered the %SD, though not to the extent observed in the Fitch-Margoliash trees. Several Wagner trees had approximately the same low %SD when the multiple addition criterion (Swofford, 1981) was used, and an example is presented in Fig. 6. Differences between the Wagner trees centered around arrangements within the *Clelia-Alsophis-Helicops* group, as in the Fitch-Margoliash trees, and in the placement of the elapid genera relative to one another.

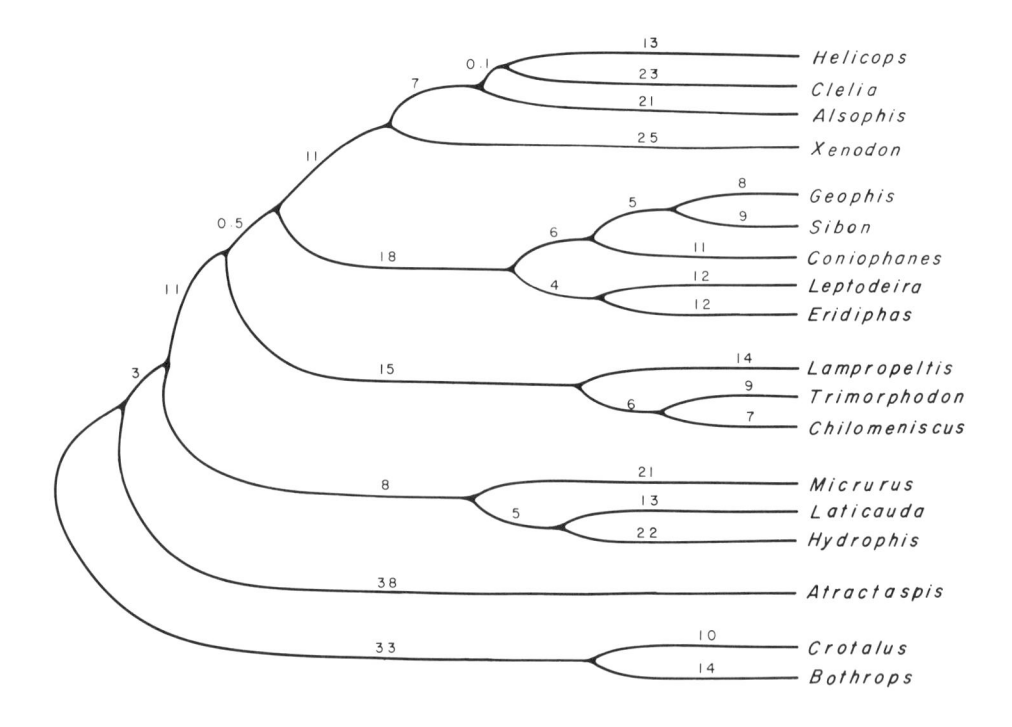

FIG. 5. A Fitch-Margoliash tree based on the corrected reciprocal matrix of MC'F albumin immunological distances (Table 3) and with branch lengths optimized using NETOPT. This tree differs from Fig. 4 only in the placement of *Atractaspis* outside the common elapid lineage and has the same standard deviation (10.1%) as the tree in Fig. 4.

The correspondence between the branching sequences seen in the best Fitch-Margoliash and Wagner trees (Figs. 4-6) is striking, and the following consistent patterns are observed: (1) the placement of *Atractaspis* as a lineage independent of other groups of advanced snakes; (2) the monophyly of the elapids and of the colubrids relative to other snakes included in the matrix; (3) the sister group relationship of the elapids and colubrids; and (4) the delineation of three major colubrid clades (*Clelia-Alsophis-Helicops-Xenodon; Geophis-Sibon-Coniophanes-Leptodeira-Eridiphas;* and *Lampropeltis-Trimorphodon-Chilomeniscus*), whose branching order differs among the best trees, although in all trees examined the branches between these major clades were very short.

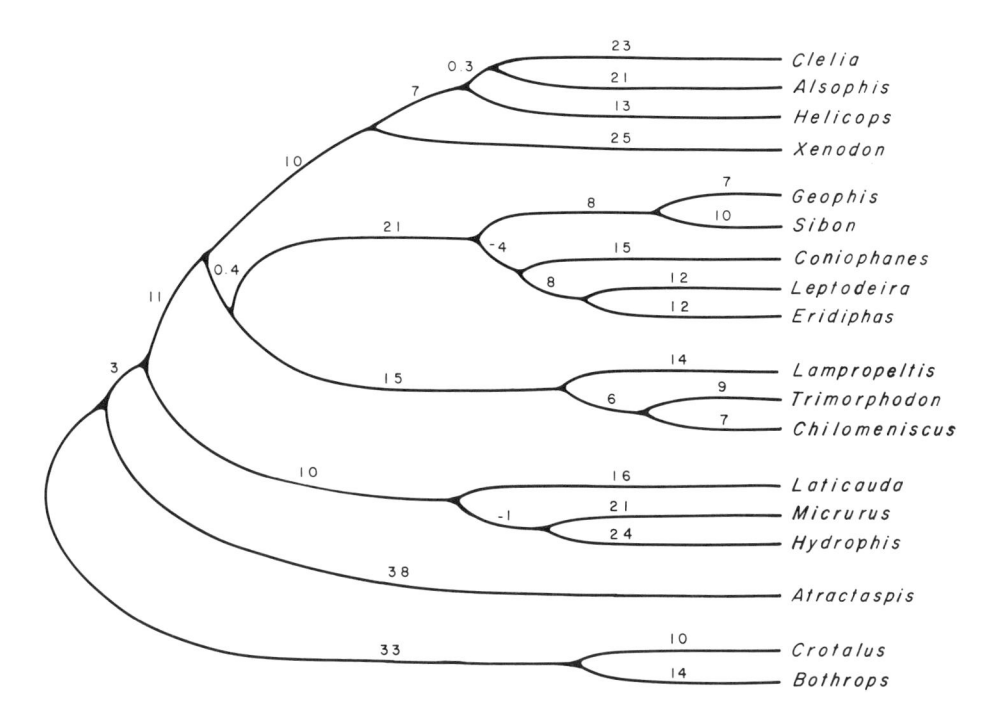

FIG. 6. A Wagner tree constructed from the corrected reciprocal matrix (Table 3) and with branch lengths optimized using NETOPT. The standard deviation is 10.6%, and the tree has two negative branches.

Only two major differences distinguish the optimized best Wagner and Fitch-Margoliash trees: (1) within the *Geophis*-etc. clade, an association is indicated between *Coniophanes* and *Leptodeira-Eridiphas* in the Wagner trees, but between *Coniophanes* and *Geophis-Sibon* in the Fitch-Margoliash trees; and (2) within the elapids, a closer association between *Micrurus* and *Hydrophis* is seen in the Wagner trees but between the latter and *Laticauda* in the Fitch-Margoliash trees. Both of the arrangements in the Wagner trees produce negative branches in the optimized Wagner trees (Fig. 6), whereas there are no negative branches in the optimized Fitch-Margoliash trees. Because of these negative branches, and the fact that the %SD's of the optimized Fitch-Margoliash trees are lower than those of the optimized Wagner trees, I consider the Fitch-Margoliash trees to be more consistent with the available molecular data. Thus the consensus phylogenetic hypothesis for these advanced snake genera (Fig. 7) shows three ambiguities in branching order among lineages: (1) in the placement of *Atractaspis*; (2) in the branching order of the three major colubrid lineages; and (3) in the branching order among *Alsophis-Helicops-Clelia*.

One-way Tests Between Atractaspis *and Other Advanced Snakes*. Results of one-way tests comparing anti-*Atractaspis bibroni* albumin to a variety of elapid and viperid taxa are presented in Table 5. (Similar values to colubrid taxa are given in the reciprocal matrix, Table 2.) Here we observe that the albumin immunological distances to viperids (81-136 units, $\bar{x} = 107$) are in general greater than those to elapids (66-106 units, $\bar{x} = 81$) or to colubrids (90-121 units, $\bar{x} = 99$). These data suggest a marginally greater similarity between the albumins of *Atractaspis* and elapids than between *Atractaspis* and either the viperids or colubrids examined.

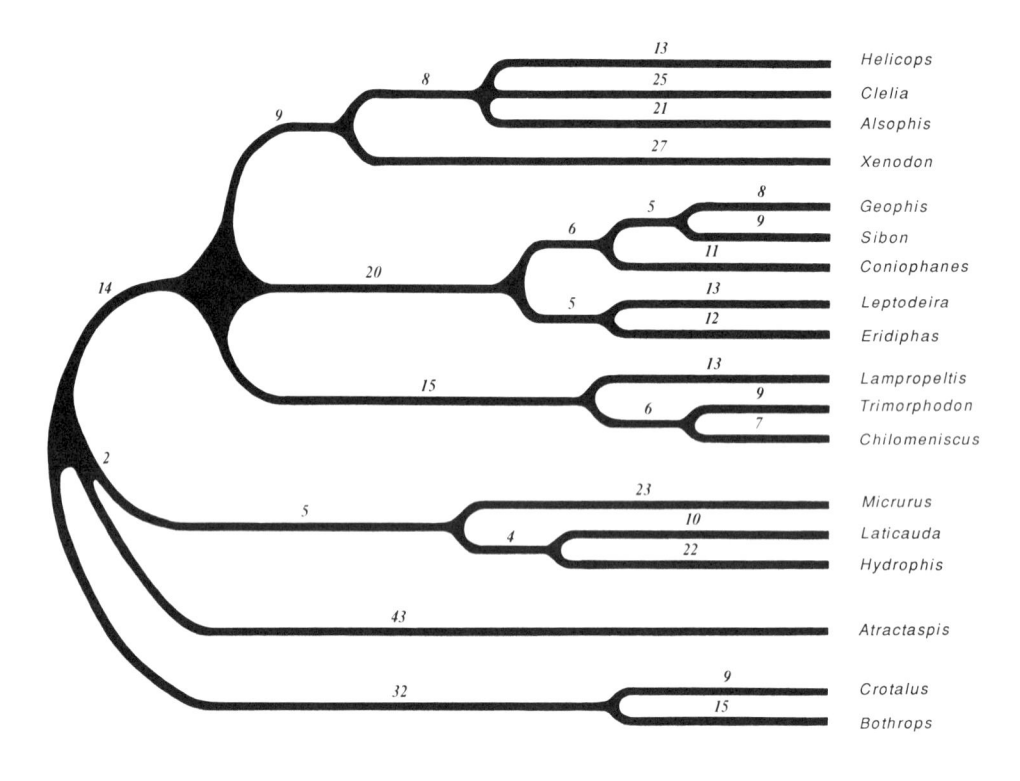

FIG. 7. A consensus phylogenetic tree for relationships among 18 genera of advanced snakes. This tree collapses into trichotomies those nodes at which the branching order of lineages differed among the Fitch-Margoliash and Wagner trees of lowest %SD (points A and B in Fig. 4). *Atractaspis* is shown separating at the base of the elapid clade, but may well separate somewhat earlier from a common elapid-colubrid lineage (see text).

Rate Tests

Rate variation in albumin evolution between families of advanced snakes, as assessed by comparisons to anti-*Boa* albumin, are given in Table 6. These data show that the albumin immunological distances to viperids (97-133 units) are less than those seen in all representatives of other families tested (143-192 units). Within the viperids, the immunological distances between *Boa* and the pit vipers are, in general, less than those to true vipers. The albumin immunological distances between *Boa* and elapids-*Atractaspis* average somewhat higher than those to colubrids, but it would be desirable to increase the sample sizes for these groups before making a statement concerning major differences among them. The reciprocal comparisons among advanced snakes (Table 2), and the phylogenetic trees constructed from these data (Figs. 4-7) using *Crotalus* and *Bothrops* as outgroups to elapids and colubrids suggest, if anything, less change in the elapids included in these comparisons (*Micrurus, Laticauda, Hydrophis*). Because the distances between *Boa* and elapids-colubrids-*Atractaspis* approach the upper limit of molecular differences detectable by MC'F, there is the possibility that rate heterogeneity would be difficult to detect in this range.

Rate variation within the colubrids was examined by inspecting the amounts of albumin change indicated along the lineages in Figs. 4-7, and by using a pool of 7 elapid antisera in MC'F reactions with representatives of several colubrid lineages (Table 7). Both methods indicate a slight conservatism in the colubrine lineage relative to the two xenodontine lineages, and other taxa (e.g., *Farancia* and *Helicops*) have changed relatively more or less than the other non-xenodontine colubrids (see also Table 2).

There is one discrepancy between the two methods in their assessment of rates. The reciprocal comparisons (Figs. 4-7) do not suggest any differences in rate between the two xenodontine clades, with about 35-40 units of change occurring along the average lineage in each. The one-way comparisons using the anti-elapid pool (Table 7) suggest that less change has occurred in the South American xenodontines (+3 to -19 units relative to *Philodryas viridissimus*) than in Central American xenodontines (+11 to +17 units relative to *P. viridissimus*). The two methods of assessing rates are not directly comparable since they involve both unidirectional and reciprocal tests, and it is not possible at this point to decide which indication is correct. Nevertheless, with the exception of some taxa (such as *Helicops* and the colubrines in general) that are conservative by both methods; and others (such as *Farancia* and possibly *Sibon-Leptodeira*) that appear fast in unidirectional tests, overall rates of change do not appear to be grossly dissimilar among the colubrid lineages examined. Similar variation has been reported in other albumin data sets (Sarich and Cronin, 1976; Cronin and Sarich, 1980; Sarich, 1973).

Table 5. MC'F immunological distances between the albumins of *Atractaspis* and representative viperids and elapids.[a]

Atractaspis			
	Atractaspis dahomeyensis	16	
	Atractaspis microlepidota	32	
Viperids			
	Crotalus enyo	81*	
	Bothrops atrox	89*	
	Sistrurus catenatus	91	
	Sistrurus miliarius	104	
	Agkistrodon piscivorus	86	
	Agkistrodon contortrix	82	
	Agkistrodon bilineatus	100	
	Bitis nasicornis	(136)	
	Bitis arietans	132	$\bar{x} = 107$
	Causus resimus	117	
	Causus maculatus	113	
	Echis ocellatus	122	
	Echis coloratus	119	
	Vipera palestinae	126	
	Pseudocerastes fieldii	95	
	Cerastes cerastes	112	
Elapids			
	Micrurus spixii	74*	
	Laticauda semifasciata	66*	
	Hydrophis melanosoma	71*	
	Dendroaspis polylepis	84*	$\bar{x} = 81$
	Elapsoidea semiannulata	89*	
	Naja haje	106*	
	Bungarus fasciatus	76*	

[a] These are one-way comparisons using an antiserum to *Atractaspis bibroni* albumin, with these exceptions: those marked with an asterisk (*) are the average of reciprocal measurements; the value in parentheses is the distance between *Bitis nasicornis* and *A. bibroni*, using an antiserum to *B. nasicornis* albumin. For comparison to these values, the average AID between *A. bibroni* and several colubrids (Table 2) is 99 units.

Table 6. Rate tests (albumin immunological distances) comparing anti-*Boa* albumin to representatives of various advanced snake families.

Viperids

  Pit vipers

Crotalus viridis	102
Crotalus enyo	99
Lachesis muta	103
Agkistrodon bilineatus	118
Bothrops atrox	97
Sistrurus miliarius	107

  True vipers

Vipera aspis	124
Bitis nasicornis	133
Echis coloratus	113
Echis ocellatus	115

Elapids and *Atractaspis*

Micrurus spixii	192
Laticauda semifasciata	192
Naja haje	183
Elapsoidea semiannulata	179
Atractaspis microlepidota	179

Colubrids

Coniophanes fissidens	151
Trimorphodon biscutatus	158
Clelia scytalina	177
Leptodeira septentrionalis	152
Sibon nebulatus	143
Xenodon severus	188
Lampropeltis getulus	179

Table 7. Rate tests comparing members of various colubrid lineages to a pool of 7 elapid antisera (*Micrurus, Hydrophis, Laticauda, Bungarus, Naja, Dendroaspis*, and *Elapsoidea*).[a]

Colubrines	*Dispholidus typus*	-19
	Thrasops jacksoni	-24
	Thelotornis capensis	-21
	Telescopus semiannulatus	-8
	Lampropeltis getulus	-8
	Trimorphodon biscutatus	-12
	Chironius carinatus	-8
	Tantilla melanocephala	-8
	Pseustes poecilonotus	-17
	Masticophis flagellum	-20
	Philothamnus angolensis	-9
	Dasypeltis scabra	0
South American xenodontines	*Clelia scytalina*	-10
	Alsophis cantherigerus	-10
	Oxyrhopus melanogenys	-14
	Xenodon severus	+3
	Helicops pastazae	-19
	Philodryas viridissimus	0
	Apostolepis assimilis	-10
	Hydrodynastes gigas	-15
Central American xenodontines	*Coniophanes fissidens*	+11
	Leptodeira septentrionalis	+17
	Eridiphas slevini	+14
	Sibon nebulatus	+17
Other lineages	*Heterodon platyrhinos*	-7
	Contia tenuis	-9
	Diadophis punctatus	-4
	Carphophis amoenus	+7
	Farancia abacura	+25
	Boaedon fuliginosus	-14
	Mehelya crossi	-3
	Rhamphiophis oxyrhynchus	+5
	Amblyodipsas polylepis	+2

[a] Rates are expressed in immunological distance units relative to *Philodryas viridissimus* = 0. Designation of the xenodontine lineages is discussed in Cadle (1982; 1984a,b).

DISCUSSION

Monophyly of the Advanced Snakes

The radioimmunoassay analysis (Table 1; Figs. 2 and 3) indicates that all advanced snakes included here are monophyletic relative to *Boa*. However, the common lineage for advanced snakes in Fig. 3 is very short, and further molecular work should include a broader array of primitive snakes in the comparisons.

Few morphological features have been found that unite the advanced snakes (Caenophidia, Colubroidea) relative to other groups of snakes, even though several workers have addressed the problem (Underwood, 1967; McDowell, 1975; Rieppel, 1979; Groombridge, 1979). Underwood discovered only two characters which in his view separated advanced from more primitive snakes: (1) the presence of only a left common carotid artery, and (2) intercostal arteries that arise from the aorta throughout the length of the trunk at intervals of several body segments (as opposed to nearly every body segment in primitive snakes). McDowell noted that a right common carotid artery persists in some viperids, but did not comment on the intercostal arteries. See Groombridge (1984) for further discussion of the systematic significance of the carotid arteries in snakes. McDowell defined a number of suprafamilial categories in snakes and indicated that there was no single known feature which would unambiguously separate the advanced snakes from more primitive ones. This approach also was used by Rieppel, who defined several primitive snake categories and therefore, by elimination, the advanced snakes, but he did not address defining characters of the Caenophidia. Holman (1979) noted several vertebral differences between primitive and advanced snakes, but most were subjective assessments of qualitative differences. Groombridge (1979) identified two putative synapomorphies of head musculature which characterize advanced snakes.

Hardaway and Williams (1976; see also Persky et al., 1976) identified another feature that unambiguously characterizes all advanced snakes relative to primitive ones. In all snakes the tips of the ribs bear cartilaginous projections which may be partially calcified. In all primitive snakes examined, including aniliids, uropeltids, xenopeltids, acrochordids, and boids, these costal cartilages are long, pointed, and curved to varying degrees. In the advanced snakes the structure of the cartilages is more complex, consisting of a base, an expanded tip that is wider than the rib, and lateral flanges that extend dorsally along the body of the rib.

In addition to these features, which appear to be more or less restricted to advanced snakes, there are many others which are generally characteristic of advanced snakes but appear in other snakes as well. The absence of coronoid bones, premaxillary teeth, and pelvic vestiges reflects simply the loss of structures primitively present. Other features are probably derived but are known to be present in some primitive snakes (e.g., broad ventral scutes) or are absent in a few advanced snakes (e.g., the arrangement of nine scutes on the head--absent in most vipers and sea snakes, and present in some primitive snakes. Other morphological characters common to advanced snakes are discussed by McDowell (1975) and Underwood (1967).

Phylogenetic Relationships among Major Groups of Advanced Snakes

Both the RIA and MC'F data indicate four major groups of advanced snakes among those examined: viperids, elapids, colubrids, and *Atractaspis* (Figs. 2-7). The lowest %SD trees (optimized Wagner and Fitch-Margoliash trees) from the RIA data (Fig. 2) suggest an early separation for the viperid lineage, and a closer relationship between elapids and colubrids than between either of these and viperids. Based on this analysis, I used viperids as an outgroup to the other advanced snakes in analyzing the MC'F data.

Most recent interpretations of morphological data (Kochva, 1962, 1963, 1978; Savitzky, 1978) are consistent with the molecular picture of a close elapid-colubrid relationship and I tentatively accept this hypothesis. A closer relationship between elapids and viperids than between either of these and colubrids has often been assumed (e.g., Bogert, 1943; Johnson, 1955, 1956; Marx and Rabb, 1965; Klauber, 1972; Rabb and Marx, 1973), apparently based on the view that the front-fanged venom delivery systems of elapids are ancestral to those of viperids. (Structurally they are quite different; see Kochva, 1978; Cadle, 1982b.) Otherwise, the similarities between elapids and viperids (hemipenial and vertebral structure, scale microstructure, etc.) appear to be primitive features (see Haas, 1938, 1952, 1962; Underwood, 1967) and thus are of dubious value as indicators of relationship.

The placement of *Atractaspis* is a bit more ambiguous on the basis of the molecular data, with some trees (Fig. 2a—RIA; Figs. 5, 6—MC'F) indicating its separation after the divergence of the viperid lineage but before the separation of the elapids and colubrids. Another tree (Fig. 4—MC'F) shows *Atractaspis* as an early derivative of the common elapid lineage. This arrangement was also seen in one of the Fitch-Margoliash trees from the RIA data, but the %SD of the optimized network was 6.4%, somewhat higher than those of other trees from these data (Fig. 2—SD = 5.98-6.1%). A final arrangement (Fig. 2c—RIA) was a three-way separation among the viperid, elapid-colubrid, and *Atractaspis* lineages.

Microcomplement fixation comparisons between *Atractaspis* and a larger sample of both elapids and viperids (Table 5) suggest no affinity with the latter, but perhaps a marginal association of *Atractaspis* with elapids (Cadle, 1983; Dessauer et al., 1987). The association of *Atractaspis* with elapids rather than viperids cannot be due to differential rates of albumin change among these three groups, since in rate tests to *Boa* (Table 6), *Atractaspis* and the elapids appear to have changed approximately equivalent amounts, while the viperids are conservative. Thus, if there were a viperid association, the immunological distances between viperids and *Atractaspis* should be less than the distances between *Atractaspis* and elapids. Therefore the phylogenetic arrangements shown in Figs. 2a, 4, and 5 are more probable than either that in Fig. 2c or an alternative placing *Atractaspis* as a branch of the viperid lineage.

No hypothesis thus far put forward on the relationships of *Atractaspis* is entirely consistent with the array of comparative data on its relationships (see Cadle, 1982b). I tentatively accept the hypotheses that the viperids are an outgroup to the remaining advanced snakes and that *Atractaspis* may represent a very early derivative of the elapid lineage, or an independent lineage diverging somewhat before the separation of the common elapid and colubrid lineages. This is not consistent with the traditional view that *Atractaspis* is a viperid derivative, nor with more recent hypotheses suggesting that it is related to aparallactine colubrids (Bourgeois, 1965; McDowell, 1968). I have compared the albumins of *Atractaspis* and two aparallactine genera and found no close association among these (Cadle, 1982b, 1983). One should note, however, that comparisons between *Atractaspis* and other colubrid lineages need to be made, and I would not be surprised to find these affecting the ultimate placement of this genus on the basis of the albumin immunological data. Dowling et al. (1983) reported an albumin immunological distance of 80 between *Atractaspis* and *Madagascarophis* (a lycodontine), and retained *Atractaspis* in the Colubridae on this basis. Without rate tests and comparisons to other groups, this conclusion is not justified. The possible association of *Atractaspis* with the elapid clade, indicated by the presently available rate-tested immunological data, suggests that certain similarities between *Atractaspis* and elapids in head structure and venom composition (Kochva et al., 1967; Parnas and Russell, 1967; Minton, 1968; Kochva and Wollberg, 1970; Kochva, 1978; McDowell, 1983, and personal communication; Lombard et al., 1986) merit further investigation, as do more extensive comparisons among *Atractaspis* and African colubrids.

Major Lineages Within Colubrids

All of the MC'F comparisons support the recognition of three major colubrid lineages among those genera sampled: *Helicops-Clelia-Alsophis-Xenodon*, *Geophis-Sibon-Coniophanes-Leptodeira-Eridiphas*, and *Lampropeltis-Trimorphodon-Chilomeniscus* (Figs. 4-6). No case can be made on the basis of the molecular data for associating any two of these lineages to the exclusion of the third (see above), and these are therefore shown as a trichotomy in the consensus tree (Fig. 7). All of these colubrids are New World forms which have classically been recognized as belonging to two subfamilies: *Lampropeltis-Trimorphodon-Chilomeniscus* in the Colubrinae and all others in the Xenodontinae. The molecular finding that two major lineages exist within the xenodontines is somewhat surprising in view of the long historical unity accorded this group (Cope, 1900; Dunn, 1928; Dowling, 1975; Dowling and Duellman, 1978). The morphological features which define the xenodontines are, however, arguably primitive features for colubrids (Cadle, 1982a, 1984c), and thus are not useful for defining a monophyletic assemblage within that group. The molecular results prompted a more intensive study of relationships among xenodontines (Cadle, 1984a-c).

Other molecular data are consistent with the pattern of colubrid relationships shown by the albumin immunological data, but in general are not precise enough to corroborate fully the hypothesis advanced here. Dessauer and colleagues (George and Dessauer,

1970; Schwaner and Dessauer, 1982) have shown on the basis of transferrin immunological comparisons (immunodiffusion and MC'F) that several groups within the xenodontines appear to be as distinct from one another as they are from colubrines. A qualitatively similar result was obtained by Minton and Salanitro (1972), using immunoelectrophoretic comparisons of snake serum proteins.

In contrast to the situation in xenodontines, the colubrines can be defined by several presumably derived morphological features (e.g., nonbilobed or asymmetrically bilobed hemipenes with a single sulcus spermaticus), and molecularly they appear to form a cohesive unit (George and Dessauer, 1970; Schwaner and Dessauer, 1982; Cadle, 1982a, 1984c). The association of *Trimorphodon* with this group has been questioned on morphological grounds (Duellman, 1958; Dowling, 1975), but it has been confirmed in several molecular studies (George and Dessauer, 1970; Minton and Salanitro, 1972; Schwaner and Dessauer, personal communication). Presumably the reluctance to include *Trimorphodon* with the colubrines stems from its possession of strongly enlarged and grooved rear maxillary teeth and a vertical pupil. Both of these features are rare in New World colubrines, but are found in numerous Old World forms (e.g., *Dispholidus, Thelotornis, Ahaetulla*). More extensive immunological comparisons among New and Old World colubrine albumins (Dowling et al., 1983; Cadle, unpublished data) show these snakes to be closely related worldwide (Cadle, 1984c, 1987). Albumin divergence among genera of colubrines of the Old and New World is no greater than that observed within either of the major xenodontine lineages, which are restricted to the New World.

The molecular data are not sufficient at this point to address the question of whether all colubrids form a monophyletic unit relative to other advanced snakes, and this remains a major evolutionary problem. Analysis of this question would be a considerable undertaking in view of the number of taxa that would need to be examined. My data (Cadle, 1982, 1984c) suggest that albumin divergence between several lineages of New and Old World colubrids is roughly similar, but comparisons of these to other advanced snakes need to be completed. Dowling et al. (1983) proposed an albumin phylogeny of colubrine, natricine, lycodontine, and xenodontine colubrids. There are a number of problems with that analysis, the most critical being that no outgroups were included to assess rates of change. In view of minor rate differences among some of these lineages (this study), no reliable cladistic conclusions concerning lineage relationshps can be drawn without such rate tests. Additionally, Dowling et al. assumed certain lineages to be monophyletic for purposes of analysis, which may not be the case for xenodontines (Cadle, 1982a, 1984c, 1985) and lycodontines (Schwaner and Dessauer, 1982; Dowling et al., 1983; Cadle, unpublished data). The apparent lack of derived features which define the colubrids as a clade relative to other advanced snakes, and some suggestions to the contrary (McDowell, 1987), make one uneasy about uncritically assuming the pattern of relationships among various colubrid subgroups for phylogenetic analyses.

Implications for Evolutionary Studies of Advanced Snakes

The molecular phylogenetic hypotheses developed here have direct bearing on studies of morphological evolution within advanced snakes. Broad evolutionary trends in advanced snakes have usually been addressed using phylogenetic hypotheses that have not been explicitly analyzed or defended (e.g., Marx and Rabb, 1972; Rabb and Marx, 1973; Kardong, 1979; Marx et al., 1982). These studies have generally assumed that extant colubrids can serve as models in assessing evolutionary events within elapids and viperids; by default, then, living colubrids have been regarded as primitive in these studies. The molecular phylogenies (Figs. 2-7) suggest, however, that viperids are the earliest derivative lineage of the advanced snake radiation, and that there should be derived features uniting the elapids and colubrids as sister groups (see Haas, 1938, 1952, 1962). They also indicate that the similarities noted between certain colubrids and viperids ("protovipers"; see Marx et al., 1982; Kardong 1979) must be due to convergence, and therefore that these colubrid taxa do not necessarily provide good models for the ancestral condition of the viperid venom apparatus (see Cadle, 1982b for discussion).

Homoplasy in morphology has caused some of the most difficult problems in unraveling the evolutionary history of snakes, especially in regard to venom delivery systems (Cadle and Sarich, 1981; Cadle 1982b, and their included references), but also in other aspects of snake morphology: dentition (enlargement and grooving of maxillary dentition), hemipenes (condition of the sulcus spermaticus), and external morphology associated with habitat use (scalation patterns, body proportions) (see Bogert, 1940; Underwood, 1967; Rabb and Marx, 1973). An extensive study of homoplasy is beyond the scope of this work, but on a broad scale the phylogenetic hypotheses developed here support the notion that it has been an important feature of advanced snake evolution: possibly three parallel acquisitions of front-fanged venom delivery systems (viperids, elapids, *Atractaspis*); the presence of enlarged, grooved maxillary teeth in all three colubrid lineages (character state polarity is a problem here, but extensive parallelism and/or reversal is indicated no matter which state is assumed to be primitive); and single hemipenes with a simple sulcus spermaticus (a derived condition) in at least two colubrid lineages (xenodontines and colubrines).

Future work can profitably be directed toward estimating the actual extent of parallelism and elucidating those evolutionary mechanisms responsible for parallelism in particular features (e.g., Savitzky, 1981, 1983). Because parallelism is thought to be pervasive within advanced snakes, this group may serve as a model system for investigating its role in adaptive radiation (see Wake, 1982).

TEMPORAL FRAMEWORK FOR THE ADVANCED SNAKE RADIATION

AN EVALUATION OF THE FOSSIL RECORD OF ADVANCED SNAKES

Considerations on the age of the advanced snake radiation have figured prominently in several areas of evolutionary biology, including estimates of speciation rates (Stanley, 1979:258), discussions on the role of specific adaptations in the radiation of extant taxa and faunal turnover (e.g., venom delivery systems, Savitzky, 1980; Holman, 1976c, 1979), and models for the historical biogeography of living groups (e.g., Hoffstetter, 1967a, b; Hoffstetter and Rage, 1977; Rage, 1978, 1981). By and large (Rage is an exception), these discussions have assumed a mid-Tertiary age for the origin and radiation of advanced snakes, particularly of colubrids. An evaluation of the fossil record suggests that this assumption is unwarranted. I review the fossil evidence here with particular regard to two major points: (1) minimum estimates for the age of advanced snakes and of the modern families within that group; and (2) possible sampling biases inherent in the caenophidian fossil record, and the resulting influence on interpretations of the history of the group. I will then consider the implications of these findings with regard to the molecular data.

Early Caenophidians

The view that advanced snakes are a very recent radiation is deeply entrenched in the literature. Romer (1966:135), for example, states, "Viewed broadly, snake history would appear to include . . . in the mid-Tertiary the development and rapid rise of the colubrids and poison-bearing derivatives." Until recently, the only pre-Miocene snakes thought to be related to caenophidians were *Archaeophis, Anomalophis, Pterosphenus,* and *Paleophis,* all suggested to be early lineages unrelated to modern caaenophidians. These are from marine Eocene deposits in Italy (*Archeophis* and *Anomalophis*); Europe, North America, and West Africa (*Paleophis*); and Africa, North America, and South America (*Pterosphenus*) (see Hoffstetter, 1958a, b, 1960, 1962). *Paleophis* and *Pterosphenus* are giant snakes of near-shore marine or estuarine habitats often associated with the whale *Basilosaurus* (Holman, 1979).

Hoffstetter (1962 and references therein) included the Paleophiiidae (containing the genera *Paleophis* and *Pterosphenus*) among the Henophidia. Romer (1966) included these among the varanoid lizards, but later (1968) agreed with Hoffstetter's assessment of their boid affinities. Rage (1975b) described *Nigerophis* (marine Paleocene of Niger) as having vertebral characteristics intermediate between primitive and advanced snakes, and suggested a relationship to the Paleophiidae and especially to *Archeophis*. These, along with *Anomalophis*, Rage hypothesized to be close to the basal stock of the caenophidian radiation (see also Hoffstetter, 1962), but he later (1977) placed *Nigerophis* in the Acrochordoidea as a sister group to the Caenophidia. Holman (1979), however, pointed out a number of vertebral characters which suggest booid affinities for at least *Paleophis* and *Pterosphenus*, and he therefore included the Paleophiidae in the Henophidia. Thus considerable uncertainty exists concerning the true phylogenetic relationships of the Paleophiidae, and available fossils do not permit a resolution of the problem. If these genera are indeed early caenophidians, which seems very doubtful, they were widespread in both Laurasian and Gondwanan landmasses in the early Tertiary but appear to be quite unrelated to any other known living or fossil caenophidians.

The caenophidian nature of *Russellophis*, known from Eocene deposits in France and Portugal (Rage, 1975a; Antunes and Russell, 1981), seems less questionable (Rage, 1975a; Richard Estes, personal communication). The French specimens (trunk and caudal vertebrae) come from deposits of Cuisian, Ypresian, and possibly Sparnacian ages (early Eocene), about 50-54 million years ago (Berggren et al, 1978); the Portuguese specimens are apparently slightly older (Antunes and Russell, 1981). Rage (1975a) was unable to estimate precisely the phylogenetic position of this genus, but noted that its caenophidian resemblance was much more striking than that of *Anomalophis*. As the first undoubted advanced snake to appear in the fossil record, *Russellophis* considerably extends the known temporal duration of the Caenophidia.

Thus our knowledge of the early history of caenophidians is somewhat vague. By early Eocene times, a number of large marine forms were in existence (*Nigerophis*, Paleophiidae, *Archeophis*, and *Anomalophis*); their phylogenetic affinity to caenophidians is highly questionable. *Russellophis*, more certainly a caenophidian, is now known from several European localities which considerably antedate the next oldest caenophidian-bearing deposits. Rage (1975b) inferred on the basis of thes meager data that the origin of advanced snakes is probably a Cretaceous event, given the current temporal distribution of the possible early members of the group.

Extant Caenophidian Families

The origin and radiation of the extant caenophidian families is also difficult to infer from available fossil material. Elapids (*Paleonaja*) and vipers (*Provipera* and *Bitis*) appear in the Middle and Lower Miocene, respectively, of Europe, and these earliest forms are anatomically modern in the skull and vertebral elements known (Hoffstetter, 1939, 1955, 1958c, 1962; Rage, 1975a). They are common in Miocene and Pliocene deposits after this period. In North America elapids and viperids are known from the

Upper and Middle Miocene, respectively (Holman, 1977a,b). Because the morphology of the earliest known members of these families is essentially the same as that of extant members, Hoffstetter (1962) suggested that each must have had a lengthy period of evolution, presently unknown, before these first fossils. (This assumes constancy of morphological change, which is doubtful.) He inferred from the absence of elapids in the Oligocene Quercy deposits of France, which have yielded colubrid fossils (see above), that *Paleonaja* was a Miocene immigrant from the east and thus parallelled a (presumed) African invasion from the east by *Naja*. The bird and mammal faunas associated with the European *Paleonaja* are similar in many respects to contemporaneous East Asian faunas, thus supporting this scenario (Hoffstetter, 1939). A similar detailed study of the earliest vipers has not yet appeared.

Mostly because of recent discoveries from the Oligocene Quercy phosphorites in France (Rage, 1974, 1984), the known temporal range of colubrids in the fossil record is greater than that of either elapids or viperids (the Upper Eocene colubrids [three vertebrae] discussed in Rage [1974] were later doubted [Rage, cited in Laurent, 1979:63; Rage, 1984]). Rage (1974) noted that it is often difficult or impossible to distinguish particular colubrid genera on the basis of vertebrae, but there seem to be two distinct vertebral types in the Quercy deposits. One type is representative of genera such as *Coluber*, *Coronella*, *Ptyas*, and *Malpolon*, and Rage provisionally assigns these to the genus *Coluber*. This snake (if indeed it is a single taxon) is abundant and present in several middle Oligocene deposits. Another snake represented by only two vertebrae is also from a Middle Oligocene deposit and appears to be a natricine (Rage, 1974). No absolute datings have been done on these deposits, but correlation of the associated fauna suggests an age of 30-31 million years (D.E. Savage, personal communication). The earliest North American colubrid fossil, *Texasophis galbreathi* (presumably a colubrine), appears contemporaneously with the Quercy fossils (Holman, 1984), whereas the colubrid record on other continents begins only in the Middle to Late Miocene (Rage, 1984).

Two points are noteworthy from these earliest colubrid fossils: (1) two major extant lineages of colubrids, colubrines and putative natricines, are present in the earliest record of the family; and (2) one genus is not distinguishable from genera living in the same area today. These observations suggest that by Oligocene times colubrids had diversified into several lineages, but they give us little indication of the actual time of origin of the family (see also Hoffstetter, 1962; Cadle, 1987).

To summarize, the available fossil record of advanced snakes indicates that by the Early Eocene several questionably advanced snakes were present, as well as *Russellophis*, whose affinities with this group are more certain. A gap of about 20 million years follows, with colubrids of two major lineages then appearing in the Middle Oligocene, approximately 30 million years ago. Viperids and elapids appear 10-12 million years later. Although present knowledge cannot exclude a much earlier differentiation for any of these groups, given these data as minimal estimates of their ages, it seems reasonable to postulate an origin for the Caenophidia in the Cretaceous (Rage, 1975b), and at least in the early Tertiary for extant families. Of course, documentation of these inferences will require the discovery of appropriate fossils. The paucity of early and middle Tertiary advanced snake fossils (in contrast to henophidian fossils; see below), and yet the tantalizing presence of forms such as *Russellophis, Nigerophis,* and the Oligocene colubrids, suggest that inferences concerning broad patterns of advanced snake evolution made from the fossil record as presently known must be made with caution. In addition to the simple presence of particular groups, other factors concerning the vagaries of fossil preservation influence these interpretations as well (discussed below).

Inferences from the Snake Fossil Record—a Critique

Since the North American fossil record of advanced snakes is better studied than that of most areas of the world, I here consider its general features, with particular concern for how possible sampling errors in the known record can influence the interpretation of major aspects of snake evolution in North America.

A summary of the known occurrences of genera in the Tertiary fossil record of snakes in North America, taken from Holman's (1979) review, is presented in Table 8. (Holman omits without comment the following: *Coniophis* cf. *precedens*, reported from the Upper Cretaceous of Alberta [Fox, 1975]; *Coniophis* sp. and *incerta sedis Dunnophis* sp. from the Middle Paleocene of Montana [Estes, 1976]; *Pseudocemophora antiqua* from the Middle Miocene of Wyoming [Holman, 1976d]; Elapidae [?*Micrurus*] from the Middle Pliocene of Florida [Auffenberg, 1963]. A report of an erycine boid of undetermined affinities from the early Eocene of Ellesmere Island [Estes and Hutchison, 1980], and an Oligocene colubrid [Holman, 1984], have appeared since Holman's review.) Practically all Tertiary snake-bearing deposits in North America come from the western United States, and thus we know essentially nothing about snake evolution in the rest of the continent (a few Miocene and Pliocene fossils in Florida notwithstanding).

Only a small number of snake genera are reported from the North American Tertiary, both in terms of total forms known and, more importantly, in terms of forms known from any single time period. This places limits on the confidence with which we can accept as real any apparent patterns of faunal composition through time. Relatively few new discoveries can significantly alter the relative abundance of taxa (for example, booids to colubroids) during a single time period, and might considerably alter ideas concerning the first appearance of particular groups. The occurrence of many of the Cretaceous and early Tertiary forms is recorded by few specimens—generally fewer than a dozen vertebrae, and often only one or two (e.g., Estes and Berberian, 1970). Such meager positive evidence for the presence of certain taxa gives us little confidence in interpreting negative evidence as real absences.

Table 8. Total numbers of snake genera represented in the North American Tertiary fossil record.[a]

	Henophidians	Caenophidians
Paleocene	3	0
Eocene	11	0
Oligocene	4	1
Lower Miocene	4	1
Middle Miocene	6	8
Upper Miocene	6	12
Lower Pliocene	2	7
Middle Pliocene	0	7
Upper Pliocene	1	11

[a] Data are from Holman (1979), with additions as noted in the text.

Furthermore, some recent discoveries have significantly changed our knowledge of the record of particular groups in North America. An example is the recent discovery of aniliid snakes (*Coniophis* and Aniliidae undetermined) in the Upper Cretaceous Campanian of Alberta (Fox, 1975), which extends the known fossil record of North American snakes by 20 million years (the next oldest being those of the late Cretaceous Lance Formation). Similarly, no North American Lower Miocene snakes of any type were known until Holman (1976a) reported five booid species from Nebraska, and very shortly thereafter (1976b) he reported both a colubrid and an additional boid, also from the Lower Miocene of Nebraska. Subsequent discovery of a Middle Oligocene colubrid (Holman, 1984) further extended the record of this family by nine million years, approximately a 40% increase in the known age for this group in North America. I suspect that we may be in for similar surprises as the study of lower vertebrate faunas in North America progresses. In particular, Paleocene and Oligocene lower vertebrate faunas are uncommon in North America (Romer, 1966; Estes, 1970, 1976), and those that do exist sample a relatively narrow subset of environments.

A change in sampling environments through the North American Tertiary is indicated by the inferred ecological conditions of those deposits bearing snake fossils. All known late Cretaceous and Paleocene deposits bearing lower vertebrates were probably deposited in freshwaters of a warm temperate coastal plain bordering the inland sea that covered much of central North America at that time (Estes, 1964, 1970, 1976; Estes and Berberian, 1970; Estes et al., 1969). Such depositional environments continued through the Eocene, with both swampy lacustrine and stream channel deposits represented (Hecht, 1959). In addition, snakes appear in estuarine or river mouth deposits of this age (e.g., Holman, 1977c). Later Tertiary deposits increasingly sample more upland and mesic habitats, and post-Eocene lower vertebrate faunas are quite different from those of the early Tertiary (Estes, 1970). Holman (1972) described a Lower Oligocene herpetofauna apparently associated with a humid forest environment, and several Miocene localities (Estes and Tihen, 1964; Holman, 1970, 1973) represent similar deposits or more arid grasslands (Holman, 1970). Tertiary mammalian and lizard faunas reflect a similar trend (Joseph T. Gregory and Jacques Gauthier, personal communications).

Changes in sampling environments influence faunal comparisons both between localities within a particular time period and between time periods. For example, rather different faunal compositions can be obtained through comparisons of essentially contemporaneous faunas that represent more upland versus lowland habitats (e.g., Hecht, 1959; Estes and Tihen, 1964; Estes and Berberian, 1970). Similar comparisons of faunas separated temporally, as well as by sampling environments, make it difficult to generalize from the fossil record about broad trends of snake evolution in North America. Holman (1979:225), for example, reports a "diminution of the henophidian fauna from Middle through Upper Miocene times" and attributes a rapid decline in boids after the Miocene to competition from the radiation of advanced snakes. As seen in Table 8, however, no decrease in the number of henophidians in the North American record is evident until after the Miocene (6 to 2 genera), and during the same interval there is also a nearly 50% reduction (12 to 7 genera) in the advanced snake fauna.

It would be of considerable interest to obtain samples from lowland post-Oligocene deposits (perhaps from the southeastern United States; Estes, 1970) for comparison with these earlier Tertiary deposits. I emphasize again the small sample sizes involved here and the ecological differences across which these comparisons are made. The strongest mid-Tertiary to Recent trend supported by the data in Table 8 is a gradual reduction in henophidian diversity from the Middle Miocene to the present. An even greater decline in henophidian diversity is seen between the Eocene and Oligocene when caenophidians are not yet known in the fossil record. From these data it is difficult to argue persuasively for a strong causal relationship between henophidian and caenophidian diversity during the North American Tertiary.

Gilmore (1938) and Romer (1966) note that snakes do not fossilize easily, and one may take this to mean fossilization in those types of deposits commonly sampled. The current practices of screening and anthill techniques undoubtedly will improve our sampling of squamate microfossils. It is significant, I believe, that many of the earliest occurrences of particular snake taxa are in unusual and rare types of deposits which permit the preservation of delicate organisms that might otherwise be destroyed. Particularly favorable sites seem to be those associated with karstic landscapes, which have yielded the oldest known colubrids (Quercy phosphorites in France; Rage, 1974) and elapids (La Grive-Saint-Alban fissure fillings, France; Hoffstetter, 1939); some of the earliest known North American colubrids (Thomas Farm limestone fissures; Auffenberg, 1963); and many of the known Tertiary snakes in South America (Itaborai, Brazil; Rage, 1978). These deposits often preserve other delicate vertebrates as well (e.g., owls and bats, Hoffstetter, 1939; caecilians, Estes and Wake, 1972). Such formations are rare, but significant additions to our knowledge of snake fossils and the evolutionary history of this group may come from identification and intensive study of such deposits.

Conclusions

There are major constraints on inferring details about the pattern of snake evolution from the fossil record as presently known. Nearly all of the available data, particularly for advanced snakes, come from temperate western North America and western Europe. No information is yet available from Asia or Australia, and Africa and South America are represented by a total of three Tertiary records of possible caenophidians. Given these uncertainties, I do not feel confident in drawing firm conclusions concerning evolutionary patterns in snakes on the basis of the present fossil record.

I consider the following inferences especially suspect: (1) The speciation rates calculated by Stanley (1979:258), and used to justify labeling the Cenozoic as the "Age of Snakes," are based on assuming a mid-Miocene (13 million years) origin for the colubrid radiation (or a more conservative age of 25 million years). The presence of the Oligocene colubrids and the possibility of a much earlier radiation of colubrids (see below) indicate that these speciation rates are much too high. (2) The assumed Miocene invasions of colubrids into both North (Holman, 1979) and South America (Hoffstetter, 1967a,b; Hoffstetter and Rage, 1977) are inferences based on negative evidence and thus suspect, as is the recent revision of the North American dispersal to Oligocene (Holman, 1984). See also Tihen (1964) and Estes (1970, 1976) for alternative views. (3) The available fossil caenophidians do not yet afford sufficient resolution to allow discrimination between differences in snake faunal composition due to environmental or depositional factors from those due to biological phenomena (e.g., competition, Holman, 1979). At best, the snake fossil record provides minimal estimates of the age of particular taxa, but reconstruction of much of ophidian history must still rely on the comparative study of living forms (see also Hoffstetter, 1962). In this discussion I have ignored totally the problems of assigning snake fossils to the correct lineage. These are covered more fully elsewhere (Cadle, 1987).

IMPLICATIONS OF THE MOLECULAR DATA

The fossil evidence summarized above indicates that the temporal dimensions of the advanced snake radiation are as uncertain as the cladistic aspects. Molecular data can potentially be used to generate testable hypotheses on the timing of separation of lineages, but in the absence of adequate fossil information to provide points for calibration of the molecular data, some assumptions must be made about the history of a particular group. Direct calibration of albumin molecular clocks using fossils has, in fact, been accomplished only with a few mammalian groups (carnivores, ungulates, primates; Carlson et al., 1978), resulting in an estimate of about 60 million years of separation corresponding to about 100 albumin immunological distance (AID) units. Application of this calibration to albumin comparisons between South American and Australian hylid frogs and marsupials (Maxson et al., 1975) resulted in estimates for the age of separation of the respective lineages on the two continents of 60-70 million years, consistent with some paleontological estimates (e.g., Tedford, 1974; McGowran, 1973). Other workers have assumed that this calibration obtains with a diverse array of other vertebrates: ranid, bufonid, and leptodactylid frogs (Wallace et al., 1973; Maxson et al., 1981; Maxson and Heyer, 1982), plethodontid salamanders (Wake et al., 1978; Larson et al., 1981; Maxson and Maxson, 1979), iguanid lizards (Gorman et al., 1971; Wyles, 1980), and elapid snakes (Cadle and Sarich, 1981). In general, the time estimates provided have been consistent with biogeographic and fossil evidence, although for most of these lower vertebrates the available information permits a relatively broad range of possible calibrations.

To rigorously defend any particular set of dates derived from molecular considerations, it is necessary to calibrate the molecular data (once approximate regularity of change is demonstrated) with reference to the group concerned. It is apparent, for example, from the available data on rates of albumin evolution in vertebrates, that there is a significant heterogeneity of absolute rates among groups, beyond the observed rate variation at lower taxonomic levels discussed earlier. This point was noted early in the work on primates, in which the prosimian lineages show less albumin change than the average anthropoid lineage (Sarich and Wilson, 1973). Subsequent studies on a variety of mammals suggest that major groups can differ in the absolute rates at which their albumins change, with groups such as anthropoids, bats, rodents, and artiodactyls having changed about 30% more than others such as carnivores and all of the prosimian lineages (Sarich and Cronin, 1980; Sarich, 1985; V.M. Sarich, personal communication). Other vertebrates known to exhibit rates of albumin evolution substantially different from mammals are birds and crocodilians (2-3 time slower than the rate observed in mammals; Prager et al., 1974) and turtles (W.E. Rainey, personal communication). These differences, of course, can be taken into account in calibrating a molecular clock for any particular group, providing enough points for calibration are available from other data (e.g., Sarich and Cronin, 1980, for hystricognath rodents).

It is important to attempt to calibrate the molecular data for particular groups, even if it results in simply placing upper and lower bounds on the albumin evolutionary rate within a group. This is preferable to using a standard calibration in view of the apparent absolute rate differences between groups. I make an attempt here to calibrate the rate of evolution of snake albumins, using the meager fossil information summarized above and some biogeographic considerations. This model relies upon more assumptions than are desirable, but offers some predictions that are testable with additional data.

The rate tests (Tables 6 and 7) to various advanced snakes and the phylogenetic analysis of albumin immunological data (Fig. 7), suggest that rate variation within the colubrids examined has been approximately equivalent between lineages and can thus be used in calibrating a molecular clock for this group. Indications thus far are that albumin evolution in some vipers and elapids has been somewhat more variable, and I reserve a consideration of the rate of albumin evolution in those groups until more taxa have been sampled. The conservatism already noted for viperids (see Dessauer et al., 1987, for further discussion) means that a different calibration than the one developed for colubrids will have to be used for this group.

The fossil evidence discussed above suggests that there are few divergences within advanced snakes that can be dated accurately through this approach. I therefore resort to a less direct method, using minimal estimates for divergences from the fossil record combined with some assumptions about the paleogeography of snakes.

Several basic pieces of information are necessary in developing this model. First, the presence of both colubrines and natricines in the fossil record at about 30 million years is taken as a minimal estimate for the age of diversification of modern lineages of colubrids. Thus the average of about 70 AID units between members of the major colubrid lineages represented in Figs. 4-7 is taken to represent a minimum age of 30 million years. While no natricines were included in these comparisons, other data (Cadle, 1984c; Dowling et al., 1983) suggest that they do not differ substantially in their albumin divergence from the colubrid lineages included in this analysis.

The upper limit, of course, cannot be set using available fossils, since there are essentially no indications of when the various advanced snake lineages separated from one another, or from primitive snakes. I therefore resort to a more speculative approach, using some assumptions about snake paleobiogeography. A major difficulty stems from uncertainties as to what constitute reasonable assumptions, but two observations permit some testable predictions. First is the separation of two xenodontine lineages (*Clelia-Alsophis-Helicops-Xenodon* vs. *Geophis-Sibon-Coniophanes-Leptodeira-Eridiphas;* Fig. 7) by 70 AID units. These lineages are apparently endemic to the New World (Cadle, 1984a-c, 1985). Thus I assume that ancestral members of these lineages were already present in Central or South America when their divergence took place.

The second observation concerns the present composition of the advanced snake fauna of Australia and inferences concerning the paleohistory of that region. The paleogeography of Australia is complex, but includes the following premises crucial to the arguments presented here: (1) a probably continuous land connection between Australia and South America via Antarctica from the late Cretaceous through middle Eocene at the latest (Raven and Axelrod, 1972, 1974; McGowran, 1973); and (2) rifting between Australia and East Antarctica beginning by the early Eocene, and suturing of the Australian and East Asian-Pacific plates during the middle Miocene (Whitmore, 1981, 1982; Keast, 1983). There are controversies over the extent to which exchange of terrestrial organisms between Australia and South America was possible during the early Tertiary (Woodburne and Zinsmeister, 1982), but exchange is indicated by the presence of marsupials, numerous related floral elements, and possibly hylid frogs on the two land masses (Raven and Axelrod, 1972, 1974; see Savage, 1973 and Tyler, 1979 for comments on the frogs). Available data (e.g., Tedford, 1974; Woodburne and Zinsmneister, 1982) suggest a sweepstakes route rather than an uninterrupted land connection, but precise time estimates from geological data for the separation of South American and Australian faunas (if, indeed, there was a single separation) are not possible. The discovery of a 40 million year old fossil marsupial in Antarctica (Woodburne and Zinsmeister, 1982) indicates that early Cenozoic climates in that region were mild enough to support terrestrial life, but the extent of faunal exchange between Australia and South America during the early Cenozoic is not known with certainty. Most of the paleontological data suggest that the Australian and South American faunas were isolated by the late Cretaceous or Paleocene (Tedford, 1974; Woodburne and Zinsmeister, 1982).

Advanced snakes in Australia currently comprise 8 colubrid genera (11 species) and about 26 elapid genera (about 65 species). All of the colubrid genera except *Myron* are found in southeast Asia, and I will assume that they are recent (Miocene or later) entrants to Australia. All of the Australopapuan elapids are endemic and I will further assume that their radiation is a Miocene event which took place after the suturing of the Australian and Asian plates, and that elapids entered Australia from southeast Asia (see Cadle, 1987, for further discussion).

The lack of colubrids in Australia, other than those showing close ties with southeast Asia, and their domination of the advanced snake fauna of South America, can then be used to estimate the upper bound for calibrating the molecular data. Assuming that a diversity of colubrids would be present in Australia had they been in South America at the time these two continents possibly last exchanged faunal elements (Paleocene; 55-65 million years ago), then an upper limit of about 60 million years is placed on the time of diversification of the endemic New World (xenodontine) colubrid lineages. Thus the 70 AID units separating major clades within the xenodontines represents about 60 million years of separation at the most.

Estimates for the rate of albumin evolution in colubrids thus correspond to 30-60 million years per 70 AID units, or 43-86 million years per 100 AID units. This rate of albumin evolution is on the same order as that obtained for several mammalian groups (55-60 million years per 100 AID units; Carlson et al, 1978; but see Sarich, 1985, for further discussion and modification). The upper limit for calibration of the snake molecular data is probably a better estimate than the lower bound, which essentially relies on negative evidence. Therefore I think that it is reasonable to postulate the separation of the colubrid lineages (two xenodontine lineages and the colubrines) in Figs. 4-6 as occurring in the middle or early Tertiary, perhaps 40-60 million years ago.

Several predictions follow from this model. If the assumptions about the origin of the Australian advanced snake fauna are correct, then we would expect both the elapids and endemic colubrids of that continent to be late derivatives of their respective lineages, with low amounts of molecular differentiation between Australian members and their closest Asian relatives. Albumin and transferrin immunological evidence on the position of Australian elapids within the elapid radiation supports this prediction (Cadle and Gorman, 1981; Schwaner et al., 1985; Cadle, unpublished data; H.C. Dessauer, personal communication). Furthermore, AID's between some Australian elapids and sea snakes are low (all except *Demansia* are in the range of 15-27 AID units; Cadle and Gorman, 1981), implying that they are also low among the Australian forms. More recently, Schwaner et al. (1985) demonstrated very low transferrin immunological distances among a broad sampling of Australian elapids, from which they inferred that this is a monophyletic lineage no more than 10 million years old. These molecular data then support a model of relatively recent colonization by few ancestral elapids, with subsequent radiation within the Australopapuan region. The predictions of the biogeographic model are, therefore, consistent with the available molecular data on the Australian snake fauna and with morphological studies (McDowell, 1967, 1969b, 1970), and they can be evaluated as more taxa are examined.

If the hypothesized middle or early Tertiary separation among colubrid lineages is accepted, then it follows from the phylogenetic hypothesis developed previously (Figs. 2, 3, and 7) that separation of the elapid and colubrid lineages was most likely a late Cretaceous-early Tertiary event, and the divergence of the viperids even earlier. The ages of these cladistic events cannot be estimated with any confidence, because of the conservative nature of albumin evolution in viperids and at least some elapids, and the current lack of appropriate outgroups with which to perform meaningful rate tests for these groups. Nor has it been possible to estimate adequately the amount of albumin change along the common elapid-colubrid lineage after the separation of the viperids, though it was probably relatively small, given the amount of albumin divergence seen within viperids (Cadle, unpublished data). This precludes an evaluation of the length of that period of common ancestry for elapids and colubrids. Nevertheless, the ages inferred here for the advanced snake clades are much older than in most current interpretations of the fossil evidence (e.g., Holman, 1979; Romer, 1966; Stanley, 1979; see critique of the fossil record given above), but are consistent with other interpretations (e.g., Hoffstetter, 1962; Rabb and Marx, 1973). Further molecular work may allow these estimates to be refined, especially by providing documentation as to the extent of rate variation and conservatism in albumin evolution within viperids and elapids.

LIMITATIONS OF MC'F STUDIES OF ALBUMINS FOR RESOLVING RELATIONSHIPS AMONG ADVANCED SNAKES

Based on the results of the molecular phylogenetic analysis presented above, I believe that there are limits to the ability of MC'F studies of albumin to resolve potential phylogenetic problems within the advanced snakes. The amount of albumin change along an average colubrid lineage since separation of the elapid clade is 45-50 units, of which 10-15 units belong to the common colubrid lineage (Figs. 4-7). For the elapid and viperid lineages included in those figures, the amounts of change are less (other comparisons within both of these groups suggest that those genera included in Figs. 4-6 are somewhat conservative). Since the estimation of phylogenetic relationships among living species by this method involves partitioning those amounts of change among taxa, clearly this imposes limits on the power of these analyses to fully resolve relationships as more taxa are examined. With nearly 300 genera of colubrids, for example, I would not expect a detailed resolution of relationships to be possible using albumin evolution for even a small proportion of these. Major lineages are likely to be separated by very few changes (as in Figs. 4-6), leaving branching orders unresolved in such cases (cf. Fig. 7). Similar phenomena are likely when many taxa within a particular clade are examined (see, for example, the discussion of South American xenodontines in Cadle, 1984a).

The well-studied situation in primates illuminates this problem. Primate lineages average about 55 units of albumin change (Cronin and Sarich, 1980), but there are far fewer living species within this entire order than there are living genera of colubrids. Thus the potential for resolving major patterns of relationships among extant species is far better in primates than in colubrids, although problems arise even in that group (e.g., among New World monkeys; Sarich and Cronin, 1976). I suspect that the major contribution of studies of albumin evolution in advanced snakes will be in delineating major clades, developing hypotheses concerning general patterns of relationships within and among clades, and pinpointing problems that might be better resolved using other approaches. Relationships within two major lineages of colubrids, some of the problems encountered for those groups, and possible solutions are discussed in Cadle (1984a-c).

Other problems in snake phylogeny may remain intractable to MC'F analyses using albumin, though the RIA method may provide a valuable starting point for further work. These are problems relating to broader patterns of evolution among snakes, particularly placement of the numerous primitive snake lineages and the advanced snake clade relative to one another. I have assumed in all of the analyses presented here that advanced snakes are monophyletic relative to all other snakes, and this is supported by both nonmolecular information (Underwood, 1967; McDowell, 1975; Rieppel, 1979; Groombridge, 1979) and the RIA analysis presented above. This question cannot be realistically addressed with the MC'F data, since the distances between *Boa* and most elapids and colubrids examined (Table 6) approach the resolving power of MC'F (AID's > 130). There is some question as to whether there is any meaningful phylogenetic information in such large distances (see previous discussion).

Immunodiffusion comparisons to advanced snakes using antisera to *Python, Loxocemus, Cylindrophis,* and *Exiliboa* give results similar to those using anti-*Boa* (Cadle, unpublished data; see also Dessauer et al., 1987), and they confirm the conservative nature of viperid albumins relative to other advanced snakes. They also indicate as great or greater albumin divergence within primitive snakes as there is between primitive and advanced snakes. This suggests that there will be problems in developing a molecular phylogeny for these groups without reference to information such as that provided by RIA. Thus, MC'F will probably not provide unequivocal solutions to these broad-scale phylogenetic problems. Analogous problems are encountered in assessing interordinal relationships among mammals using immunological approaches. Sarich and Cronin (1980), Cronin and Sarich (1980) and Sarich (1985) discuss these problems in more detail and suggest some ways of addressing them through molecular data.

SUMMARY AND CONCLUSIONS

1. Molecular comparisons of albumins support the hypothesis of monophyly of the advanced snakes. Four major clades are recognized: viperids, elapids (including sea snakes), colubrids, and *Atractaspis*.
2. Viperids appear to be the earliest extant derivative of the advanced snake radiation, with separation of the colubrid and elapid lineages occurring later. *Atractaspis* may be an early derivative of the elapid lineage, or it may have separated somewhat earlier from the common elapid-colubrid lineage.
3. A current major question is whether the extant colubrids form a monophyletic group relative to other advanced snakes. Neither molecular nor morphological data are currently capable of resolving this problem.
4. Xenodontine colubrids comprise at least two lineages equivalent in their albumin divergence from one another as are other major colubrid lineages.
5. Front-fanged venom delivery systems and several other aspects of the morphology of advanced snakes have been subject to parallel or convergent evolution.
6. The fossil record of advanced snakes does not support some generalizations about snake evolution that are commonly seen in the literature. These relate primarily to the overall temporal framework for the radiation. Molecular data, combined with minimal estimates from fossils for the ages of colubrid lineages and with assumptions drawn from biogrography, suggest a middle to early Tertiary radiation for the colubrid lineages examined, and therefore an earlier separation (Cretaceous or earliest Tertiary) for the elapid and viperid lineages.

Appendix A
Specimens Used for the Production of Antisera

Abbreviations for repositories of specimens are:

CAS   California Academy of Sciences, San Francisco

MVZ   Museum of Vertebrate Zoology, University of California, Berkeley

LACM   Natural History Museum of Los Angeles County

ZRCS   Zoological Reference Collection, University of Singapore

Specimens without a collection number are not yet catalogued.

Antisera to those species for which more than one specimen are indicated were produced from a pool of blood or tissue from those specimens.

HENOPHIDIA

Boidae

Boa constrictor. Mexico: Michoacán, Río Tepalcatepec at Capirio (MVZ 172374).

CAENOPHIDIA

Viperidae

Bitis nasicornis. Kenya: Kakamega District, Kakamega Forest (CAS).
Bothrops atrox. Peru: Depto. Amazonas, vicinity of Huampami, Río Cenepa (MVZ 163351).
Crotalus enyo. Mexico: Baja California, road to Bahía de los Angeles, 11 km ESE jct. with Mexico Hwy. 1 (MVZ 164980).

Elapidae

Bungarus fasciatus. Probable origin southern China, sent as a preserved blood sample by the zoology department, Hong Kong University (no voucher).

Dendroaspis polylepis. Zimbabwe: Gwanda District, vicinity of Gwanda (MVZ 176493).

Elapsoidea semiannulata. Zimbabwe: Gwanda District, vicinity of Gwanda (MVZ 176494).

Hydrophis melanosoma. Malaysia: Johore, mouth of the Muar River (MVZ 175454, + 1 voucher in ZRCS).

Laticauda semifasciata. Philippine Islands: Gato Island, north of Cebu Island (LACM 129648).

Micrurus spixi. Peru: Depto. Amazonas, vicinity of Huampami, Río Cenepa (MVZ 163329).

Naja haje. Zimbabwe: Bulawayo District, obtained from Bulawayo Snake Park (MVZ 176495).

Atractaspis

Atractaspis bibroni. Locality unknown (MVZ 175868).

Colubridae

Alsophis cantherigerus. Greater Antilles: Cayman Brac Island (CAS 151252).

Chilomeniscus cinctus. Mexico: Baja California, El Sombrero Trailer Park, La Paz (MVZ 170778-170779).

Clelia scytalina. Mexico: Oaxaca, Rte. 185, 6.1 mi. S Oaxaca-Veracruz border (MVZ 146947).

Coniophanes fissidens. Guatemala: Depto. San Marcos, Finca Santa Julia, 1.5 km E San Rafael Pié de la Cuesta (MVZ 146568-146573).

Eridiphas slevini. Mexico: Baja California, 7.5 mi. (by Mexico Hwy. 1) S of Mulegé. (MVZ 161418).

Farancia abacura. South Carolina: Aiken County, Savannah River Plant (MVZ 175901).

Geophis nasalis. Guatemala: Depto. San Marcos, Finca Santa Julia, 1.5 km E San Rafael Pié de la Cuesta (MVZ 146585-146595).

Helicops pastazae. Peru: Depto. Amazonas, vicinity of Huampami, Río Cenepa (MVZ 163276 and 163278).

Heterodon platyrhinos. Georgia: Emanuel County, Swainsboro (MVZ 175886).

Hydrodynastes gigas. Locality unknown (MVZ 175897).

Imantodes cenchoa. Mexico: Quintana Roo, 19 km NNW (by Coba road) of Tulúm (MVZ 175893).

Lampropeltis getulus. California: Sonoma County, Skaggs Springs (MVZ 175859).

Leptodeira septentrionalis. Mexico: Tamaulipas, Gomez Farias Road, 3.2 mi. W jct. Rte. 85 (MVZ 146959).

Ninia sebae. Guatemala: Depto. San Marcos, Finca Santa Julia, 1.5 km E San Rafael Pié de la Cuesta (MVZ 146736-146743).

Oxyrhopus melanogenys. Peru: Depto. Amazonas, vicinity of Huampami, Rio Cenepa (MVZ 163300).

Philodryas viridissimus. Peru: Depto. Amazonas, vicinity of Huampami, Rio Cenepa (MVZ 163308).

Sibon nebulatus. Mexico: Guerrero, 2.8 km NE (by road from Atoyac to Puerto del Gallo) town of Rio Santiago (MVZ 172373).

Trimorphodon biscutatus. Mexico Michoacán, 10 km E (by road) of Apatzingan (MVZ 170791).

Xenodon severus. Peru: Depto. Amazonas, Quebrada Caterpiza, Rio Santiago (MVZ 175339).

Appendix B
Specimens Used in Immunological Cross Reactions

Repositories for specimens are abbreviated as in Appendix A, plus the following:

NMZB   National Museums of Zimbabwe, Bulawayo

EK   Laboratory colony maintained by Dr. Elazar Kochva, Tel Aviv University

UGZD   Zoology Department reptile collection, University of Ghana

Viperidae

Agkistrodon bilineatus. Locality unknown (MVZ 172409).
Agkistrodon contortrix. North Carolina: Macon County, 2.9 mi. ESE (airline) of Highlands (MVZ 173595).
Agkistrodon piscivorus. Georgia: Emanuel County, 0.5 mi. E Georgia Rte. 56 on Old Savannah Road (MVZ 175883).
Bitis arietans. Ghana: Eastern Region, vicinity of Legon (UGZD).
Causus maculatus. Ghana: Eastern Region, Legon Hill, University of Ghana campus, Legon (MVZ 176461).
Causus resimus. Kenya: Kisumu District, Chenilil (CAS 152792).
Cerastes cerastes. Israel: exact locality unknown (EK).
Crotalus viridis. California: Napa County, 1 mi. ENE Aetna Springs on Pope Valley Road (MVZ).
Echis coloratus. Israel: exact locality unknown (EK).
Echis ocellatus. Ghana: Upper Region, Wa (MVZ).
Lachesis muta. Peru: Depto. Amazonas, vicinity of Huampami, Río Cenepa (MVZ 163372).
Pseudocerastes fieldii. Israel: exact locality unknown (EK).
Sistrurus catenatus. Texas: Parker County, 5 air miles S Aledo (MVZ 175885).
Vipera aspis. Locality unknown (MVZ 175902).
Vipera palestinae. Israel: exact locality unknown (EK).

Appendix B

Atractaspis

Atractaspis dahomeyensis. Ghana: Eastern Region, Legon Hill, University of Ghana Campus, Legon (MVZ 176456).
Atractaspis microlepidota. Somalia: Lower Juba Region, Juba Sugar Project near Mareri (CAS 153176).

Colubridae

Amblyodipsas polylepis. Zimbabwe: Gwelo District, Gwelo (MVZ 176471).
Apostolepis assimilis. Brazil: Edo. São Paulo, exact locality unknown (obtained from Instituto Butantan) (MVZ 176334).
Boaedon fuliginosus. Locality unknown (MVZ 172384).
Carphophis amoenus. Oklahoma: McCurtain County, Beaver's Bend State Park (MVZ).
Chironius carinatus. Peru: Depto. Amazonas, vicinity of Huampami, Río Cenepa (MVZ 163250).
Contia tenuis. California: Sonoma County, Skaggs Springs (MVZ).
Dasypeltis scabra. Zimbabwe: Beitbridge District, Mazunga River bridge (NMZB 6013).
Diadophis punctatus. California: Napa County, Sulfur Canyon, 2 mi. SW (by air) of St. Helena (MVZ).
Dispholidus typus. Zimbabawe: Umtali District, Umtali (MVZ 176475).
Masticophis flagellum. Mexico: Baja California, El Sombrero Trailer Park, La Paz (MVZ 170753).
Mehelya crossi. Ghana: Eastern Region, Legon Hill, University of Ghana campus, Legon (MVZ 176441).
Philothamnus angolensis. Zimbabwe: Melsetter District, Martin Forest Reserve (MVZ 176481).
Pseustes poecilonotus. Peru: Depto. Amazonas, vicinity of Huampami, Río Cenepa (MVZ 163309).
Rhamphiophis oxyrhynchus. Ghana: Upper Region, Wa Secondary School, Wa (MVZ 176504).
Tantilla melanocephala. Peru: Depto. Madre de Dios, Albergue, Río Madre de Dios approx. 12 km (airline) E of Puerto Maldonado (MVZ 173765).
Telescopus semiannulatus. Zimbabwe: Umtali District, Umtali (MVZ 176490).
Thelotornis capensis. Zimbabwe: Gwanda District, vicinity of Gwanda (MVZ 176491).
Thrasops jacksoni. Kenya: Kakamega District, Kakamega Forest (CAS 152795).

Literature Cited

Antunes, M. T., and D. E. Russell. 1981. Le gisement de Silveirinha (Bas Mondego, Portugal): la plus ancienne faune de vertébrés éocènes connue en Europe. Comp. Rend. Acad. Sci., Paris, 293:773-776.

Arnheim N., E. M. Prager, and A. C. Wilson. 1969. Immunological prediction of sequence differences among proteins: chemical comparison of chicken, quail, and pheasant lysozymes. J. Biol. Chem. 244:2085-2094.

Auffenberg, W. W. 1963. The fossil snakes of Florida. Tulane Stud. Zool. 10:131-216.

Baba, M., L. Darga, and M. Goodman. 1980. Biochemical evidence on the phylogeny of Anthropoidea. In *Evolutionary Biology of New World Monkeys and Continental Drift,* ed. R. L. Ciochon and A. B. Chiarelli, pp. 423-443. New York: Plenum Press.

Baba, M., Weiss, M. L., M. Goodman, and J. Czelusniak. 1982. The case of *Tarsier* hemoglobin. Syst. Zool. 31:156-165.

Beard, J. M., and M. Goodman. 1976. The haemoglobins of *Tarsius bancanus*. In *Molecular Anthropology*, ed. M. Goodman and R. Tashian, pp. 239- 255. New York: Plenum Press.

Benjamin, D. C., J. A. Berzofsky, I. J. East, F. R. N. Gurd, C. Hannum, S. J. Leach, E. Margoliash, J. G. Michael, A. Miller, E. M. Prager, M. Reichlin, E. E. Sercarz, S. J. Smith-Gill, P. E. Todd, and A. C. Wilson. 1984. The antigenic structure of proteins: A reappraisal. Ann. Rev. Immunol. 2:67-101.

Literature Cited

Beverly, S., and A. C. Wilson. 1982. Molecular evolution in *Drosophila* and higher diptera. I. Microcomplement fixation studies of a larval hemolymph protein. J. Mol. Evol. 18:251-264.

Berggren, W. A., M. C. McKenna, J. Hardenbol, and J. D. Obradovich. 1978. Revised Paleogene polarity time scale. J. Geol. 86:67-81.

Bogert, C. M. 1940. Herpetological results of the Vernay Angola expedition. Part I. Snakes, including an arrangement of African Colubridae. Bull. Amer. Mus. Nat. Hist. 77:1-107.

_____. 1943. Dentitional phenomena in cobras and other elapids with notes on adaptive modifications of fangs. Bull. Amer. Mus. Nat. Hist. 81:285-360.

Bourgeois, M. 1965. Contribution à la morphologie comparée du crâne des ophidiens de l'Afrique Centrale. Publs. Univ. Off. Congo, Lubumbashi, 28:1-293.

Bruce, E. J., and F. J. Ayala. 1979. Phylogenetic relationships between man and the apes: electrophoretic evidence. Evolution 33:1040-1056.

Cadle, J. E. 1982a. "Evolutionary relationships among advanced snakes." Ph.D. dissertation, University of California, Berkeley.

_____. 1982b. Problems and approaches in the interpretation of the evolutionary history of venomous snakes. Mem. Inst. Butantan 46:255- 274.

_____. 1983. Phylogenetic relationships of African back-stabbing snakes, genus *Atractaspis*. Abstr. SSAR/HL meeting, University of Utah, p. 49.

_____. 1984a. Molecular systematics of xenodontine colubrid snakes. I. South American xenodontines. Herpetologica 40:8-20.

_____. 1984b. Molecular systematics of xenodontine colubrid snakes. II. Central American xenodontines. Herpetologica 40:21-30.

_____. 1984c. Molecular systematics of xenodontine colubrid snakes. III. Overview of xenodontine phylogeny and the history of New World snakes. Copeia 1984:641-652.

_____. 1985. The neotropical colubrid snake fauna: lineage components and biogeography. Syst. Zool. 34:1-20.

_____. 1987. The geographic distribution of snakes. In *Snakes: Ecology and Evolutionary Biology,* ed. R. A. Seigel, J. T. Collins, and S. S. Novak, pp. 77-105. New York: Macmillan Publishing Co.

Cadle, J. E., and G. C. Gorman. 1981. Albumin immunological evidence and the relationships of sea snakes. J. Herpetol. 15:329-334.

Cadle, J. E., and V. M. Sarich. 1981. An immunological assessment of the phylogenetic position of New World coral snakes. J. Zool., London, 195:157-167.

Carlson, S. S., A. C. Wilson, and R. D. Maxson. 1978. Do albumin clocks run on time? Science 200:1183-1185.

Champion, A. B., E. M. Prager, D. Wachter, and A. C. Wilson. 1974. Microcomplement fixation. In *Biochemical and Immunological Taxonomy of Animals,* ed. C. A. Wright, pp. 397-416. London, Academic Press.

Champion, A. B., K. L. Soderberg, and A. C. Wilson. 1975. Immunological comparison of azurins of known amino acid sequence. J. Mol. Evol. 5:291-305.

Cope, E. D. 1900. The crocodilians, lizards, and snakes of North America. Rep. U.S. Nat. Mus. 1898:153-1249.

Coulter, A. R., R. D. Harris, and S. K. Sutherland. 1981. Enzyme immunoassay and radioimmunoassay: their use in the study of Australian and exotic snake venoms. In *Proc. Melbourne Herpetological Symposium,* ed. C. A. Banks and A. A. Martin, pp. 39-43.

Cronin, J. E., and W. E. Meikle. 1979. The phyletic position of *Theropithecus:* congruence among molecular, morphological, and paleontological evidence. Syst. Zool. 28:259-269.

Cronin, J. E., and V. M. Sarich. 1980. Tupaiid and archonta phylogeny: the macromolecular evidence. In *Comparative Biology and Evolutionary Relationships of Tree Shrews*, ed. W. P. Luckett, pp. 239-312. New York: Plenum Press.

De Jong, W. W., and M. Goodman. 1982. Mammalian phylogeny studied by sequence analysis of the eye lens protein α-crystallin. Z. f. Saugetierkunde 47:257-276.

Dene, H., M. Goodman, M. C. McKenna, and A. E. Romero-Herrera. 1982. *Ochotona princeps* (pika) myoglobin: an appraisal of lagomorph phylogeny. Proc. Nat. Acad. Sci. USA 79:1917-1920.

Densmore, L. D. 1981. "Biochemical and immunological systematics of the order Crocodilia." Ph.D. dissertation, Louisiana State University.

Dessauer, H. C., J. E. Cadle, and R. Lawson. 1987. Patterns of snake evolution suggested by their proteins. Fieldiana: Zoology, new series 34:1-34.

Dowling, H. G. 1975. The Nearctic snake fauna. In *1974 Yearbook of Herpetology*, ed. H. G. Dowling, pp. 191-202. New York: American Museum of Natural History.

Dowling, H. G., and W. E. Duellman. 1978. *Systematic herpetology: a synopsis of families and higher categories*. New York: Herpetological Information Search Systems publications.

Dowling, H. G., R. Highton, G. C. Maha, and L. R. Maxson. 1983. Biochemical evaluation of colubrid snake phylogeny. J. Zool., London, 201:309-329.

Duellman, W. E. 1958. A monographic study of the colubrid snake genus *Leptodeira*. Bull. Amer. Mus. Nat. Hist. 114:1-152.

Dunn, E. R. 1928. A tentative key and arrangement of the American genera of colubridae. Bull. Antivenin Inst. Amer. 2:18-24.

Estes, R. 1964. Fossil vertebrates from the late Cretaceous Lance Formation, Eastern Wyoming. Univ. Calif. Publ. Geol. Sci. 49:1-180.

_____. 1970. Origin of the recent North American lower vertebrate fauna: an inquiry into the fossil record. Forma et Functio 3:139-163.

_____. 1976. Middle Paleocene lower vertebrates from the Tongue River formation, Southeastern Montana. J. Paleontol. 50:500-520.

Estes, R., and P. Berberian. 1970. Paleoecology of a late Cretaceous vertebrate community from Montana. Breviora no. 343:1-35.

Estes, R., P. Berberian, and C. Meszoely. 1969. Lower vertebrates from the late Cretaceous Hell Creek Formation, McCone County, Montana. Breviora no. 337:1-33.

Estes, R., and J. H. Hutchison. 1980. Eocene lower vertebrates from Ellesmere Island, Canadian Arctic archipelago. Palaeog., Palaeoclim., Palaeoecol. 30:325-347.

Estes, R., and J. A. Tihen. 1964. Lower vertebrates from the Valentine formation of Nebraska. Amer. Midl. Nat. 72:453-472.

Estes, R., and M. H. Wake. 1972. The first fossil record of caecilian amphibians. Nature 239:228-231.

Farris, J. S. 1972. Estimating phylogenetic trees from distance matrices. Amer. Nat. 106:645-668.

_____. 1981. Distance data in phylogenetic analysis. In *Advances in Cladistics*, Proc. 1st Meeting of the Willi Hennig Soc., ed. V. A. Funk and D. R. Brooks, pp. 3-23. New York: New York Botanical Gardens.

_____. 1985. Distance data revisited. Cladistics 1:67-85.

Farris, J. S., A. G. Kluge, and M. F. Mickevich. 1979. Paraphyly of the *Rana boylii* species group. Syst. Zool. 28:627-634.

Felsenstein, J. 1978. The number of evolutionary trees. Syst. Zool. 27:27-33.

_____. 1982. Numerical methods for inferring evolutionary trees. Quart. Rev. Biol. 57:379-404.

_____. 1984. Distance methods for inferring phylogenies: a justification. Evolution 38:16-24.

Ferris, S. D., A. C. Wilson, and W. M. Brown. 1981. Evolutionary tree for apes and humans based on cleavage maps of mitochondrial DNA. Proc. Nat. Acad. Sci. USA 78:2432-2436.

Fitch, W. M. 1976. Molecular evolutionary clocks. In *Molecular Evolution*, ed. F. J. Ayala, pp. 160-178. Sunderland, Mass.: Sinauer Assoc.

_____. 1979. Cautionary remarks on using gene expression events in parsimony procedures. Syst. Zool. 28:375-379.

_____. 1982. The challenges to Darwinism since the last centennial and the impact of molecular studies. Evolution 36:1133-1143.

Fitch, W. M., and C. Langley. 1976. Evolutionary rates in proteins: neutral mutations and the molecular clock. In *Molecular Anthropology*, ed. M. Goodman and R. Tashian, pp. 197-219. New York: Plenum Press.

Fitch, W. M., and E. Margoliash. 1967. Construction of phylogenetic trees. Science 155:279-284.

Fox, R. C. 1975. Fossil snakes from the Upper Milk River Formation (Upper Cretaceous), Alberta. Can. J. Earth Sci. 12:1557-1563.

Friday, A. E. 1980. The status of immunological distance data in the construction of phylogenetic classifications: a critique. In *Chemosystematics: Principles and Practice,* Syst. Assoc. spec. vol. 16, ed. F. A. Bisby, J. G. Vaughan, and C. A. Wright, pp. 289-304. London: Academic Press.

George, D. W., and H. C. Dessauer. 1970. Immunological correspondence of transferrins and the relationships of colubrid snakes. Comp. Biochem. Physiol. 33:617-627.

Gilmore, C. W. 1938. Fossil snakes of North America. Geol. Soc. Amer. spec. pap. 9:1-96.

Goodman, M., J. Czelusniak, G. W. Moore, A. E. Romero-Herrera, and G. Matsuda. 1979. Fitting the gene lineage into its species lineage, a parsimony strategy illustrated by cladograms constructed from globin sequences. Syst. Zool. 28:132-163.

Gorman, G. C., D. G. Buth, and J. S. Wyles. 1980. *Anolis* lizards of the eastern Caribbean: a case study in evolution. III. A cladistic analysis of albumin immunological data, and the definition of species groups. Syst. Zool. 29:143-158.

Gorman, G. C., A. C. Wilson, and M. Nakanishi. 1971. A biochemical approach towards the study of reptilian phylogeny: evolution of serum albumin and lactic dehydrogenase. Syst. Zool. 20:167-186.

Groombridge, B. C. 1979. Variations in morphology of the superficial palate of henophidian snakes and some possible systematic implications. J. Nat. Hist. 13:447-475.

————. 1984. The facial carotid artery in snakes (Reptilia: Serpentes): Variations and possible cladistic significance. Amphib.-Reptilia 5:145-155.

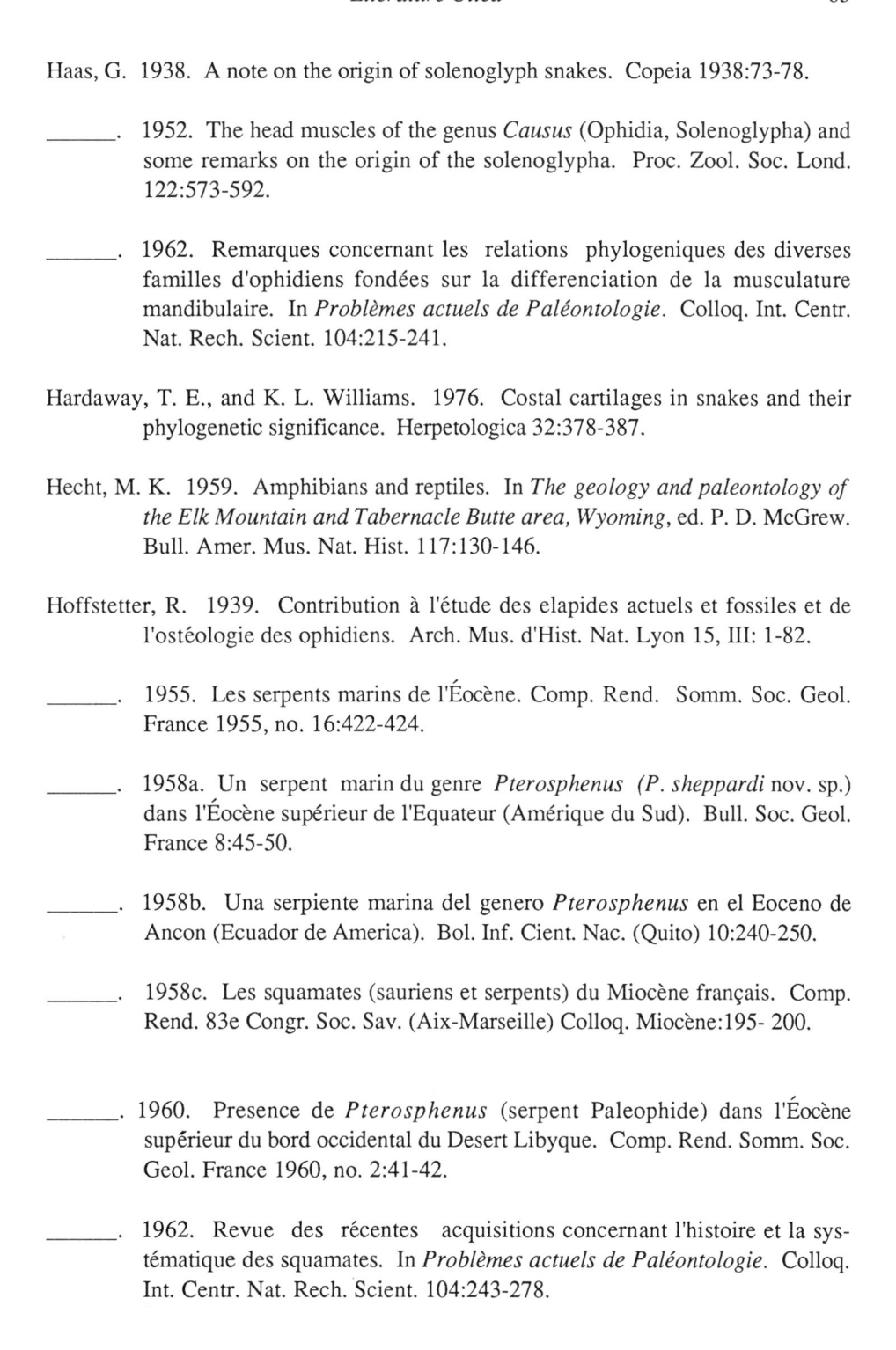

Haas, G. 1938. A note on the origin of solenoglyph snakes. Copeia 1938:73-78.

———. 1952. The head muscles of the genus *Causus* (Ophidia, Solenoglypha) and some remarks on the origin of the solenoglypha. Proc. Zool. Soc. Lond. 122:573-592.

———. 1962. Remarques concernant les relations phylogeniques des diverses familles d'ophidiens fondées sur la differenciation de la musculature mandibulaire. In *Problèmes actuels de Paléontologie*. Colloq. Int. Centr. Nat. Rech. Scient. 104:215-241.

Hardaway, T. E., and K. L. Williams. 1976. Costal cartilages in snakes and their phylogenetic significance. Herpetologica 32:378-387.

Hecht, M. K. 1959. Amphibians and reptiles. In *The geology and paleontology of the Elk Mountain and Tabernacle Butte area, Wyoming*, ed. P. D. McGrew. Bull. Amer. Mus. Nat. Hist. 117:130-146.

Hoffstetter, R. 1939. Contribution à l'étude des elapides actuels et fossiles et de l'ostéologie des ophidiens. Arch. Mus. d'Hist. Nat. Lyon 15, III: 1-82.

———. 1955. Les serpents marins de l'Éocène. Comp. Rend. Somm. Soc. Geol. France 1955, no. 16:422-424.

———. 1958a. Un serpent marin du genre *Pterosphenus (P. sheppardi* nov. sp.) dans l'Éocène supérieur de l'Equateur (Amérique du Sud). Bull. Soc. Geol. France 8:45-50.

———. 1958b. Una serpiente marina del genero *Pterosphenus* en el Eoceno de Ancon (Ecuador de America). Bol. Inf. Cient. Nac. (Quito) 10:240-250.

———. 1958c. Les squamates (sauriens et serpents) du Miocène français. Comp. Rend. 83e Congr. Soc. Sav. (Aix-Marseille) Colloq. Miocène:195- 200.

———. 1960. Presence de *Pterosphenus* (serpent Paleophide) dans l'Éocène supérieur du bord occidental du Desert Libyque. Comp. Rend. Somm. Soc. Geol. France 1960, no. 2:41-42.

———. 1962. Revue des récentes acquisitions concernant l'histoire et la systématique des squamates. In *Problèmes actuels de Paléontologie*. Colloq. Int. Centr. Nat. Rech. Scient. 104:243-278.

_____. 1967a. Observations additionelles sur les serpents du Miocène de Colombie et rectification concernant la date d'arrivée des Colubrides en Amérique du Sud. Comp. Rend. Somm. Soc. Geol. France 1967:209-210.

_____. 1967b. Remarques sur les dates d'implantation de differents groupes de serpents terrestres en Amérique du Sud. Comp. Rend. Somm. Soc. Geol. France 1967:93-94.

_____. 1968. Nuapua, un gisement de vertébrés pleistocènes dans le Chaco Bolivien. Bull. Mus. Nat. Hist. Nat. Zool., Paris, 40:823-836.

_____. 1971. Los vertebrados Cenozoicos de Colombia: yacimientos, faunas, problemas planteados. Geol. Colombiana 8:37-62.

Hoffstetter, R., and J.-C. Rage. 1977. Le gisement de vertébrés Miocènes de La Venta (Colombie) et sa faune de serpents. Ann. Paléontologie (Vertébrés) 63:161-190.

Holman, J. A. 1970. Herpetofauna of the Wood Mountain Formation (Upper Miocene) of Saskatchewan. Can. J. Earth Sci. 7:1317-1325.

_____. 1972. Herpetofauna of the Calf Creek local fauna (Lower Oligocene: Cypress Hills Formation) of Saskatchewan. Can. J. Earth Sci. 9:1612-1631.

_____. 1973. Reptiles of the Egelhoff local fauna (Upper Miocene) of Nebraska. Contr. Mus. Paleontol. Univ. Michigan 24:125-134.

_____. 1976a. Snakes of the Gering Formation (Lower Miocene) of Nebraska. Herpetologica 32:88-94.

_____. 1976b. A boid and a colubrid snake from the Harrison Formation (Lower Miocene: Arikareean) of Sioux County, Nebraska. Herpetologica 32: 387-389.

_____. 1976c. Snakes and stratigraphy. Mich. Academician 8:387-396.

_____. 1976d. Snakes of the Split Rock Formation (Middle Miocene, Central Wyoming. Herpetologica 32:419-426.

_____. 1977a. Upper Miocene snakes (Reptilia, Serpentes) from southeastern Nebraska. J. Herpetol. 11:323-335.

_____. 1977b. Amphibians and reptiles from the Gulf Coast Miocene of Texas. Herpetologica 33:391-403.

_____. 1977c. Upper Eocene snakes (Reptilia, Serpentes) from Georgia. J. Herpetol. 11:141-145.

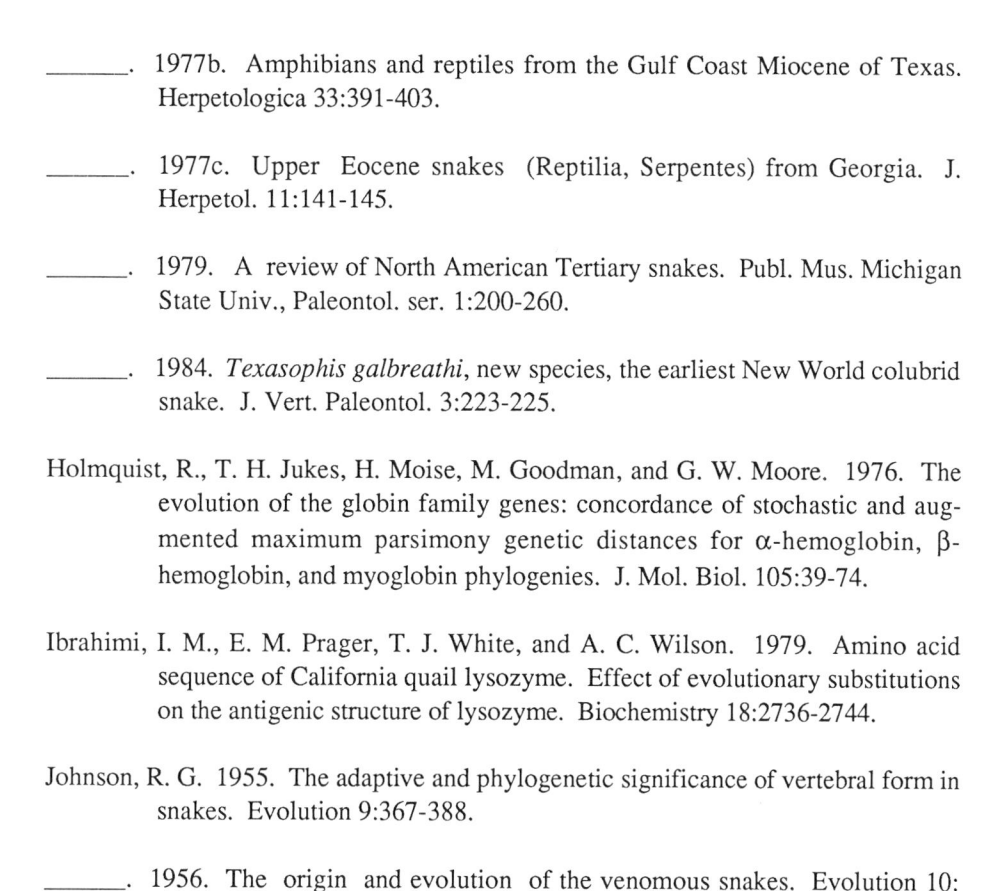

_____. 1979. A review of North American Tertiary snakes. Publ. Mus. Michigan State Univ., Paleontol. ser. 1:200-260.

_____. 1984. *Texasophis galbreathi*, new species, the earliest New World colubrid snake. J. Vert. Paleontol. 3:223-225.

Holmquist, R., T. H. Jukes, H. Moise, M. Goodman, and G. W. Moore. 1976. The evolution of the globin family genes: concordance of stochastic and augmented maximum parsimony genetic distances for α-hemoglobin, β-hemoglobin, and myoglobin phylogenies. J. Mol. Biol. 105:39-74.

Ibrahimi, I. M., E. M. Prager, T. J. White, and A. C. Wilson. 1979. Amino acid sequence of California quail lysozyme. Effect of evolutionary substitutions on the antigenic structure of lysozyme. Biochemistry 18:2736-2744.

Johnson, R. G. 1955. The adaptive and phylogenetic significance of vertebral form in snakes. Evolution 9:367-388.

_____. 1956. The origin and evolution of the venomous snakes. Evolution 10:55-65.

Jolles, J., I. M. Ibrahimi, E. M. Prager, F. Schoentgen, P. Jolles, and A. C. Wilson. 1979. Amino acid sequence of pheasant lysozyme: evolutionary change affecting processing of prelysozyme. Biochemistry 13:2744-2752.

Jolles, J., Schoentgen, F., P. Jolles, E. M. Prager, and A. C. Wilson. 1976. Amino acid sequence and immunological properties of chachalaca egg white lysozyme. J. Mol. Evol. 8:59-78.

Kardong, K. V. 1979. "Protovipers" and the evolution of snake fangs. Evolution 33:433-443.

Keast, J. A. 1983. In the steps of Alfred Russell Wallace: biogeography of the Asian-Australian interchange zone. In *Evolution, Time and Space: The Emergence of the Biosphere,* Syst. Assoc. spec. vol. 23, ed. R. W. Sims, J. H. Price, and P. Whalley, pp. 367-407. London: Academic Press.

King, M.-C., and A. C. Wilson. 1975. Evolution at two levels in humans and chimpanzees. Science 188:107-116.

Klauber, L. M. 1972. *Rattlesnakes: Their Habits, Life Histories, and Influence on Mankind.* Berkeley and Los Angeles: University of California Press.

Kochva, E. 1962. On the lateral jaw musculature of the solenoglypha with remarks on some other snakes. J. Morph. 110:227-284.

———. 1963. Development of the venom gland and trigeminal muscles in *Vipera palestinae*. Acta Anatomica 52:49-89.

———. 1978. Oral glands of the reptilia. In *Biology of the Reptilia*, ed. C. Gans, 8:43-161. London: Academic Press.

Kochva, E., and M. Wollberg. 1970. The salivary glands of the Aparallactinae (colubridae) and the venom glands of *Elaps* (Elapidae) in relation to the taxonomic status of this genus. Zool. J. Linn. Soc. 49:217-224.

Kochva, E., M. Wollberg, and R. Sobol. 1967. The special pattern of the venom gland in *Atractaspis* and its bearing on the taxonomic status of the genus. Copeia 1967:763-772.

Langley, C. H., and Fitch, W. M.. 1974. An examination of the constancy of the rate of molecular evolution. J. Mol. Evol. 3:161-177.

Larson, A., D. B. Wake, L. R. Maxson, and R. Highton. 1981. A molecular phylogenetic perspective on the origins of morphological novelties in the salamanders of the tribe plethodontini (Amphibia, Plethodontidae). Evolution 35:405-422.

Laurent, R. F. 1979. Herpetofaunal relationships between Africa and South America. In *The South American herpetofauna: Its Origin, Evolution, and Dispersal*, ed. W. E. Duellman, Monogr. Mus. Nat. Hist., Univ. Kansas 7:55-71.

Levine, L. 1978. Micro-complement fixation. In *Handbook of Experimental Immunology,* ed. D. M. Weir, pp. B5.1-B5.9. Oxford: Blackwell Scientific.

Lombard, R. E., H. Marx, and G. B. Rabb. 1986. Morphometrics of the ectopterygoid in advanced snakes (Colubroidea): a concordance of shape and phylogeny. Biol. J. Linn. Soc. 27:133-164.

Lowenstein, J. M. 1980a. Immunospecificity of fossil collagens. In *Biogeochemistry of Amino Acids*, ed. P. E. Hare, pp. 277-308. New York: Wiley.

———. 1980b. Species-specific proteins in fossils. Naturwissenschaften 67: 343-346.

———. 1981. Immunological reactions from fossil material. Phil. Trans. Roy. Soc. Lond. B292:143-149.

Lowenstein, J. M., V. M. Sarich, and B. J. Richardson. 1981. Albumin systematics of the extinct mammoth and Tasmanian wolf. Nature 291:409-411.

Maeda, N., and W. M. Fitch. 1981. Amino acid sequence of a myoglobin from Lace monitor lizard, *Varanus varius,* and its evolutionary implications. J. Biol. Chem. 256:4301-4309.

Malnate, E. V. 1960. Systematic division and evolution of the colubrid snake genus *Natrix*, with comments on the subfamily Natricinae. Proc. Acad. Nat. Sci. Philadelphia 112:41-71.

Mao, S.-H., B. Y. Chen, and H. M. Chang. 1983. Immunotaxonomic relationships of sea snakes to terrestrial elapids. Comp. Biochem. Physiol. 74A:869-872.

Mao, S.-H., and H. C. Dessauer. 1971. Selectively neutral mutations, transferrins, and the evolution of natricine snakes. Comp. Biochem. Physiol. 40A: 669-680.

Margoliash, E.A., A. Nisonoff, and M. Reichlin. 1970. Immunological activity of cytochrome *c*. J. Biol. Chem. 245:931-945.

Marx, H., and G. B. Rabb.. 1965. Relationships and zoogeography of the viperine snakes (Family Viperidae). Fieldiana, Zool. 44:161-206.

———. 1972. Phyletic analysis of fifty characters of advanced snakes. Fieldiana, Zool 63:1-321.

Marx, H., G. B. Rabb, and S. J. Arnold. 1982. *Pythonodipsas* and *Spalerosophis*, colubrid snake genera convergent to the vipers. Copeia 1982:553-561.

Maxson, L. R., and W. R. Heyer. 1982. Leptodactylid frogs and the Brazilian shield: an old and continuing adaptive radiation. Biotropica 14:10-15.

Maxson, L. R., R. Highton, and D. B. Wake. 1979. Albumin evolution and its phylogenetic implications in the plethodontid salamander genera *Plethodon* and *Ensatina*. Copeia 1979:502-508.

Maxson, L. R., and R. D. Maxson. 1979. Comparative albumin and biochemical evolution in salamanders. Evolution 33:1057-1062.

Maxson, L. R., V. M. Sarich, and A. C. Wilson. 1975. Continental drift and the use of albumin as an evolutionary clock. Nature 255:397-400.

Maxson, L. R., A.-R. Song, and R. Lopata. 1981. Phylogenetic relationships among North American toads, genus *Bufo*. Biochem. Syst. Ecol. 9:347-350.

Maxson, L. R., and A. C. Wilson. 1975. Albumin evolution and organismal evolution in tree frogs (Hylidae). Syst. Zool. 24:1-15.

McCarthy, C. 1985. Monophyly of elapid snakes (Serpentes: Elapidae). An assessment of the evidence. Zool. J. Linn. Soc. 83:79-93.

McDowell, S. B. 1967. *Aspidomorphus*, a genus of New Guinea snakes of the family Elapidae, with notes on related genera. J. Zool., London 151:497-543.

_____. 1968. Affinities of the snakes usually called *Elaps lacteus* and *E. dorsalis*. Zool. J. Linn. Soc. 47:561:561-578.

_____. 1969a. Notes on the Australian sea-snake *Ephalophis greyi* M. Smith (Serpentes: Elapidae: Hydrophiinae) and the origin and classification of sea-snakes. Zool. J. Linn. Soc. 48:333-349.

_____. 1969b. *Toxicocalamus*, a New Guinea genus of snakes of the family Elapidae. J. Zool., London 159:443-511.

_____. 1970. On the status and relationships of the Solomon Island elapid snakes. J. Zool., London 161:145-190.

_____. 1975. A catalogue of the snakes of New Guinea and the Solomons, with special reference to those in the Bernice P. Bishop Museum. Part II. Anilioidea and Pythoninae. J. Herpetol. 9:1-79.

_____. 1983. The cervicomandibularis muscle of colubroid snakes. Abstr. SSAR/HL meeting, University of Utah, p. 78.

_____. 1987. Snake systematics. In *Snakes: Ecology and Evolutionary Biology,* ed. R. A. Seigel, J. T. Collins, and S. S. Novak, pp. 3-50. New York: Macmillan Publishing Co.

McGowran, B. 1973. Rifting and drift of Australia and the migration of mammals. Science 180:759-761.

Minton, S. A. 1968. Antigenic relationships of the venom of *Atractaspis microlepidota* to that of other snakes. Toxicon 6:59-64.

Minton, S. A., and M. S. da Costa. 1975. Serological relationships of sea snakes and their evolutionary implications. In *The Biology of Sea Snakes,* ed. W. A. Dunson, pp. 33-55. Baltimore: University Park Press.

Minton, S. A., and S. K. Salanitro. 1972. Serological relationships among some colubrid snakes. Copeia 1972:246-252.

Moore, C. W., M. Goodman, C. Callahan, R. Holmquist, and H. Moise. 1976. Stochastic versus augmented maximum parsimony method for estimating superimposed mutations in the divergent evolution of protein sequences: methods tested on cytochrome c amino acid sequences. J. Mol. Biol. 105:15-37.

Nakanishi, M., A. C. Wilson, R. A. Nolan, G. C. Gorman, and G. S. Bailey. 1969. Phenoxyethanol: protein preservative for taxonomists. Science 163:681-683.

Parnas, I., and F. E. Russell. 1967. Effects of venom on nerve, muscle, and neuromuscular junction. In *Animal Toxins,* ed. F. E. Russell and P. R. Saunders, pp. 401-415. Oxford: Pergamon Press.

Patton, J. L., and M. F. Smith. 1981. Molecular evolution in *Thomomys*: phyletic systematics, paraphyly, and rates of evolution. J. Mammal. 62:493-500.

Peacock, D. 1981. Data handling for phylogenetic trees. In *Biochemical Evolution,* ed. H. Gutfreund, pp. 88-115. Cambridge: Cambridge University Press.

Persky, B., H. M. Smith, and K. L. Williams. 1976. Additional observations on ophidian costal cartilages. Herpetologica 32:399-401.

Post, T., and T. M. Uzzell. 1981. The relationships of *Rana sylvatica* and the monophyly of the *Rana boylii* group. Syst. Zool. 30:170-180.

Prager, E. M., A. H. Brush, R. A. Nolan, M. Nakanishi, and A. C. Wilson. 1974. Slow evolution of transferrin and albumin in birds according to microcomplement fixation analysis. J. Mol. Evol. 3:243-262.

Prager, E. M., G. W. Welling, and A. C. Wilson. 1978. Comparison of various immunological methods for distinguishing among mammalian pancreatic ribonucleases of known amino acid sequence. J. Mol. Evol. 10:293-307.

Prager, E. M., and A. C. Wilson. 1971a. The dependence of immunological cross-reactivity upon sequence resemblance among lysozymes. I. Microcomplement fixation studies. J. Biol. Chem. 246:5978-5989.

_____. 1971b. The dependence of immunological cross-reactivity upon sequence resemblance among lysozymes. II. Comparison of precipitin and microcomplement fixation results. J. Biol. Chem. 246:7010-7017.

_____. 1976. Congruency of phylogenies derived from different proteins. A molecular analysis of the phylogenetic position of cracid birds. J. Mol. Evol. 9:45-57.

Rabb, G. B., and H. Marx. 1973. Major ecological and geographic patterns in the evolution of colubroid snakes. Evolution 27:69-83.

Radinsky, L. 1978. Do albumin clocks run on time? Science 200:1182-1183.

Rage, J.-C. 1974. Les serpents des phosphorites du Quercy. Palaeovertebrata 6:274-303.

_____. 1975 a. Un caenophidien primitif (Reptilia, Serpentes) dans l'Éocène inférieur. Comp. Rend. Somm. Soc. Geol. France 2:46-47.

_____. 1975 b. Un serpent du Paléocène du Niger: étude preliminaire sur l'origine des caenophidiens (Reptilia, Serpentes). Comp. Rend. Acad. Sci., Paris, 281:515-518.

_____. 1977. L'origine des Colubroides et des Acrochordoides (Reptilia, Serpentes). Comp. Rend. Acad. Sci., Paris, ser. D, 286:595-597.

_____. 1978. Une connexion continentale entre Amérique du Nord et Amérique du Sud au Crétacé supérieur? l'exemple des vertébrés continentaux. Comp. Rend. Somm. Soc. Geol. France 1978:281-185.

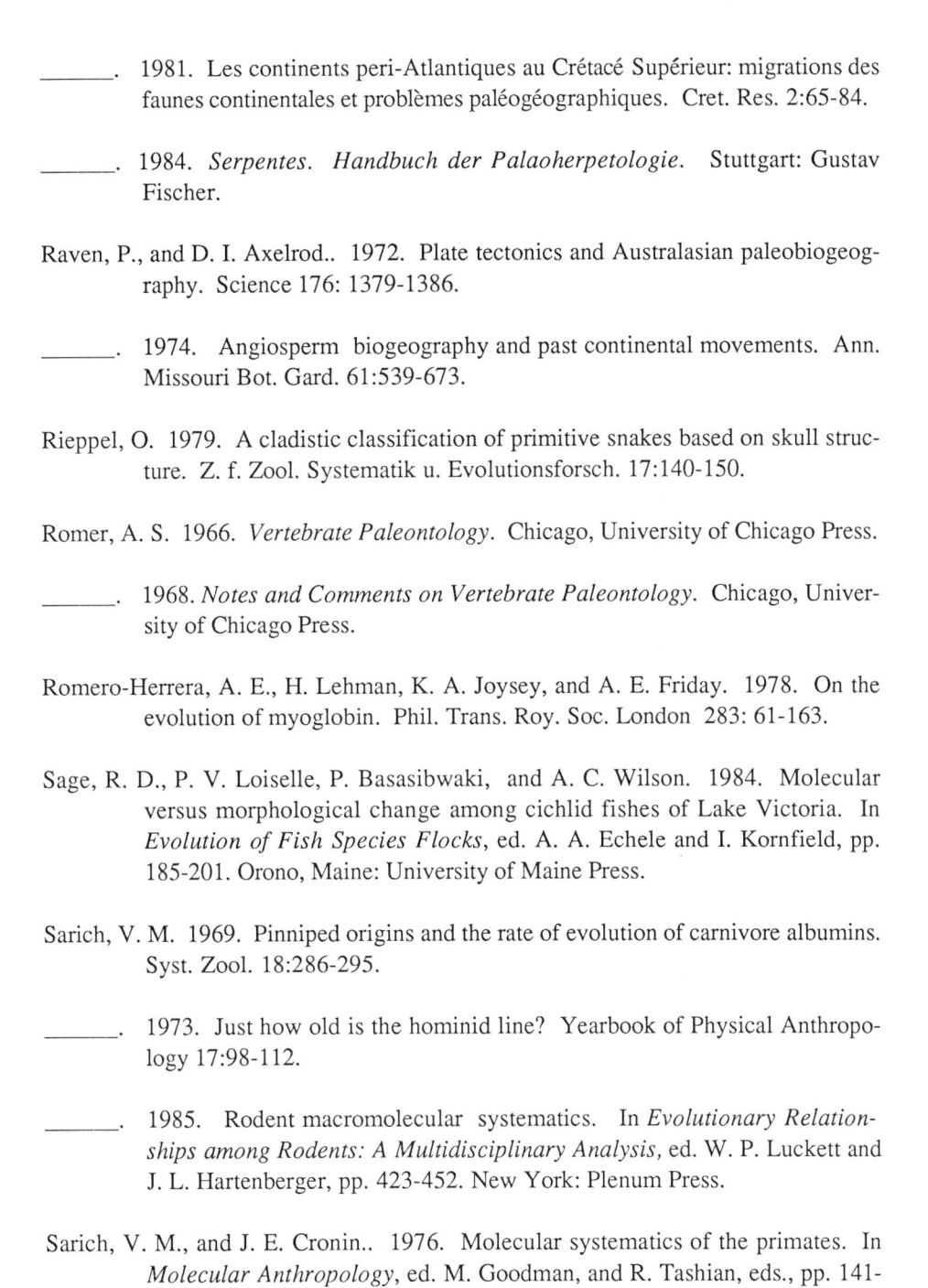

_____. 1981. Les continents peri-Atlantiques au Crétacé Supérieur: migrations des faunes continentales et problèmes paléogéographiques. Cret. Res. 2:65-84.

_____. 1984. *Serpentes. Handbuch der Palaoherpetologie.* Stuttgart: Gustav Fischer.

Raven, P., and D. I. Axelrod.. 1972. Plate tectonics and Australasian paleobiogeography. Science 176: 1379-1386.

_____. 1974. Angiosperm biogeography and past continental movements. Ann. Missouri Bot. Gard. 61:539-673.

Rieppel, O. 1979. A cladistic classification of primitive snakes based on skull structure. Z. f. Zool. Systematik u. Evolutionsforsch. 17:140-150.

Romer, A. S. 1966. *Vertebrate Paleontology.* Chicago, University of Chicago Press.

_____. 1968. *Notes and Comments on Vertebrate Paleontology.* Chicago, University of Chicago Press.

Romero-Herrera, A. E., H. Lehman, K. A. Joysey, and A. E. Friday. 1978. On the evolution of myoglobin. Phil. Trans. Roy. Soc. London 283: 61-163.

Sage, R. D., P. V. Loiselle, P. Basasibwaki, and A. C. Wilson. 1984. Molecular versus morphological change among cichlid fishes of Lake Victoria. In *Evolution of Fish Species Flocks,* ed. A. A. Echele and I. Kornfield, pp. 185-201. Orono, Maine: University of Maine Press.

Sarich, V. M. 1969. Pinniped origins and the rate of evolution of carnivore albumins. Syst. Zool. 18:286-295.

_____. 1973. Just how old is the hominid line? Yearbook of Physical Anthropology 17:98-112.

_____. 1985. Rodent macromolecular systematics. In *Evolutionary Relationships among Rodents: A Multidisciplinary Analysis,* ed. W. P. Luckett and J. L. Hartenberger, pp. 423-452. New York: Plenum Press.

Sarich, V. M., and J. E. Cronin.. 1976. Molecular systematics of the primates. In *Molecular Anthropology,* ed. M. Goodman, and R. Tashian, eds., pp. 141-170. New York: Plenum Press.

———. 1980. South American mammal molecular systematics, evolutionary clocks, and continental drift. In *Evolutionary Biology of New World Monkeys and Continental Drift,* ed. R. L. Ciochon and A. B. Chiarelli, pp. 399-421. New York: Plenum Press.

Sarich, V. M., and A. C. Wilson. 1967a. Rates of albumin evolution in primates. Proc. Natl. Acad. Sci. USA:142-148.

———. 1967b. Immunological time scale for hominid evolution. Science 158: 1200-1203.

———. 1973. Generation time and genomic evolution in primates. Science 179: 1144-1147.

Savage, J. M. 1973. The geographical distribution of frogs: patterns and predictions. In *Evolutionary Biology of the Anurans*, ed. J. L. Vial, pp. 353-445. Columbia: University of Missouri Press.

Savitzky, A. H. 1978. "The origin of the New World proteroglyphous snakes and its bearing on the study of venom delivery systems in snakes." Ph.D. dissertation, University of Kansas.

———. 1980. The role of venom delivery strategies in snake evolution. Evolution 34:1194-1204.

———. 1981. Hinged teeth in snakes: an adaptation for swallowing hard-bodied prey. Science 212: 346-349.

———. 1983. Coadapted character complexes among snakes: fossoriality, piscivory and durophagy. Amer. Zool. 23:397-409.

Scanlan, D., L. R. Maxson, and W. E. Duellman.. 1980. Albumin evolution in marsupial frogs. Evolution 34:222-229.

Schwaner, T. D., and H. C. Dessauer. 1982. Comparative immunodiffusion survey of snake transferrins focused on the relationships of the natricines. Copeia 1982:541-549.

Schwaner, T. D., P. R. Baverstock, H. C. Dessauer, and G. A. Mengden. 1985. Immunological evidence for the phylogenetic relationships of Australian elapid snakes. In *Biology of Australasian Frogs and Reptiles,* ed. G. Grigg, R. Shine, and H. Ehmann, pp. 177-184. New South Wales: Royal Zoological Society.

Selander, R. K. 1982. Phylogeny. In *Perspectives on Evolution*, ed. R. Milkman, pp. 32-59. Sunderland, Mass.: Sinauer Assoc.

Shochat, D., and H. C. Dessauer. 1981. Comparative immunological study of the albumins of *Anolis* of the Caribbean Islands. Comp. Biochem. Physiol. 68A:67-73.

Sibley, C. G., and J. A. Ahlquist. 1984. The phylogeny of the hominoid primates, as indicated by DNA-DNA hybridization. J. Mol. Evol. 20:2-15.

Stanley, S. M. 1979. *Macroevolution, Pattern and Process*. San Francisco, W. H. Freeman.

Swofford, D. L. 1981. On the utility of the distance Wagner procedure. In *Advances in Cladistics; Proc. 1st Meeting of the Willi Hennig Soc.,* ed. V. A. Funk and D. R. Brooks, pp. 25-43. New York: New York Botanical Gardens.

———. 1982a. "Wagner Procedure Program (WAGPROC) documentation, version 3." Available from the author.

———. 1982b. "Network Optimization Program (NETOPT) documentation, version 4." Available from the author.

Tedford, R. H. 1974. Marsupials and the new paleogeography. In *Paleogeographic provinces and provinciality,* ed. C. A. Ross, Soc. Econ. Paleont. Mineral. spec. publ. 21:109-126.

Throckmorton, L. H. 1978. Molecular phylogenetics. In *Biosystematics in Agriculture*. Beltsville Symposia in Agricultural Research, vol. 2, ed. J. A. Romberger, R. H. Foote, L. Knutson, and P. L. Lentz, pp. 221-239. New York: Wiley.

Tihen, J. A. 1964. Tertiary changes in the herpetofaunas of temperate North America. Senck. Biol. 45: 265-279.

Tyler, M. J. 1979. Herpetofaunal relationships of South America with Australia. In *The South American Herpetofauna: Its origin, evolution, and dispersal*. Monogr. Mus. Nat. Hist., University of Kansas, 7:73- 106, ed. W. E. Duellman.

Underwood, G. 1967. *A Contribution to the Classification of Snakes.* London: British Museum (Natural History).

Voris, H. K. 1977. A phylogeny of the sea snakes (Hydrophiidae). Fieldiana: Zoology 70:79-169.

Wake, D. B. 1981. The application of allozyme evidence to problems in the evolution of morphology. In *Evolution Today*, Proc. 2nd Int. Congr. Syst. Evol. Biol., ed. G. E. Scudder and J. L. Reveal, pp. 257-270.

———. 1982. Functional and evolutionary morphology. Perspectives in Biol. and Med. 25:603-620.

Wake, D. B., L. R. Maxson, and G. Z. Wurst. 1978. Genetic differentiation, albumin evolution, and their biogeographic implications in plethodontid salamanders of California and southern Europe. Evolution 32:529-539.

Wallace, D., M.-C. King, and A. C. Wilson. 1973. Albumin differences among ranid frogs: taxonomic and phylogenetic implications. Syst. Zool. 22:1-13.

Wallace, D., and A. C. Wilson. 1972. Comparison of frog albumins with those of other vertebrates. J. Mol. Evol. 2:72-86.

White, T. J., I. M. Ibrahimi, and A. C. Wilson. 1978. Evolutionary substitutions and the antigenic structure of globular proteins. Nature 274:92-94.

Whitmore, T. C. 1981. *Wallace's Line and Plate Tectonics.* Oxford: Oxford University Press.

———. 1982. Wallace's Line: a result of plate tectonics. Ann. Missouri Bot. Gard. 69:668-675.

Wiley, E. O. 1981. *Phylogenetics, The Theory and Practice of Phylogenetic Systematics.* New York: Wiley.

Wilson, A. C., S. S. Carlson, and T. J. White. 1977. Biochemical evolution. Ann. Rev. Biochem. 46:573-639.

Wilson, A. C., and V. M. Sarich. 1969. A molecular time scale for human evolution. Proc. Nat. Acad. Sci. USA 63:1088-1093.

Woodburne, M. O., and W. J. Zinsmeister. 1982. Fossil land mammal from Antarctica. Science 218:284-286.

Wright, S. 1978. *Variability Within and Among Natural Populations.* Evolution and the Genetics of Populations, Vol. 4. Chicago: University of Chicago Press.

Wyles, J. S. 1980. "Phylogenetic studies of iguanid lizards." Ph.D. dissertation, University of California, Los Angeles.

Wyss, A. R., M. J. Novacek, and M. C. McKenna. 1987. Amino acid sequence versus morphological data and the interordinal relationships of mammals. Mol. Biol. Evol. 4:99-116.

Yang, S. Y., and J. L. Patton. 1981. Genic variability and differentiation in the Galapagos finches. The Auk 98:230-242.

FORSYTH LIBRARY
FORT HAYS STATE UNIVERSITY